Functional Oxide Based
Thin-Film Materials

Functional Oxide Based Thin-Film Materials

Special Issue Editor
Dong-Sing Wuu

MDPI • Basel • Beijing • Wuhan • Barcelona • Belgrade • Manchester • Tokyo • Cluj • Tianjin

Special Issue Editor
Dong-Sing Wuu
National Chung Hsing University
Taiwan

Editorial Office
MDPI
St. Alban-Anlage 66
4052 Basel, Switzerland

This is a reprint of articles from the Special Issue published online in the open access journal *Crystals* (ISSN 2073-4352) (available at: https://www.mdpi.com/journal/crystals/special_issues/Zinc_Oxide).

For citation purposes, cite each article independently as indicated on the article page online and as indicated below:

LastName, A.A.; LastName, B.B.; LastName, C.C. Article Title. *Journal Name* **Year**, *Article Number*, Page Range.

ISBN 978-3-03928-837-3 (Pbk)
ISBN 978-3-03928-838-0 (PDF)

© 2020 by the authors. Articles in this book are Open Access and distributed under the Creative Commons Attribution (CC BY) license, which allows users to download, copy and build upon published articles, as long as the author and publisher are properly credited, which ensures maximum dissemination and a wider impact of our publications.

The book as a whole is distributed by MDPI under the terms and conditions of the Creative Commons license CC BY-NC-ND.

Contents

About the Special Issue Editor vii

Preface to "Functional Oxide Based Thin-Film Materials" ix

Dong-Sing Wuu
Special Issue Editorial: Functional Oxide Based Thin-Film Materials
Reprinted from: *Crystals* **2020**, *10*, 195, doi:10.3390/cryst10030195 5 1

Theopolina Amakali, Likius. S. Daniel, Veikko Uahengo, Nelson Y. Dzade and Nora H. de Leeuw
Structural and Optical Properties of ZnO Thin Films Prepared by Molecular Precursor and Sol–Gel Methods
Reprinted from: *Crystals* **2020**, *10*, 132, doi:10.3390/cryst10020132 5

Che-Yuan Yeh, Yi-Man Zhao, Hui Li, Fei-Peng Yu, Sam Zhang and Dong-Sing Wuu
Growth and Photocatalytic Properties of Gallium Oxide Films Using Chemical Bath Deposition
Reprinted from: *Crystals* **2019**, *9*, 564, doi:10.3390/cryst9110564 17

Shang-Chou Chang and Huang-Tian Chan
Effect of Nitrogen Flow in Hydrogen/Nitrogen Plasma Annealing on Aluminum-Doped Zinc Oxide/Tin-Doped Indium Oxide Bilayer Films Applied in Low Emissivity Glass
Reprinted from: *Crystals* **2019**, *9*, 310, doi:10.3390/cryst9060310 27

Jiahong Zheng, Runmei Zhang, Kangkang Cheng, Ziqi Xu, Pengfei Yu, Xingang Wang and Shifeng Niu
Research on the High-Performance Electrochemical Energy Storage of a NiO@ZnO (NZO) Hybrid Based on Growth Time
Reprinted from: *Crystals* **2019**, *9*, 47, doi:10.3390/cryst9010047 35

Yi-Yun Chen, Yuan-Chang Jhang, Chia-Jung Wu, Hsiang Chen, Yung-Sen Lin and Chia-Feng Lin
Fabrication of GaOx Confinement Structure for InGaN Light Emitter Applications
Reprinted from: *Crystals* **2018**, *8*, 418, doi:10.3390/cryst8110418 49

Víctor Herrera, Tomás Díaz-Becerril, Eric Reyes-Cervantes, Godofredo García-Salgado, Reina Galeazzi, Crisóforo Morales, Enrique Rosendo, Antonio Coyopol, Román Romano and Fabiola G. Nieto-Caballero
Highly Visible Photoluminescence from Ta-Doped Structures of ZnO Films Grown by HFCVD
Reprinted from: *Crystals* **2018**, *8*, 395, doi:10.3390/cryst8100395 55

Huangpu Han, Bingxi Xiang and Jiali Zhang
Simulation and Analysis of Single-Mode Microring Resonators in Lithium Niobate Thin Films
Reprinted from: *Crystals* **2018**, *8*, 342, doi:10.3390/cryst8090342 73

Xuguang Luo, Yao Li, Hong Yang, Yuanlan Liang, Kaiyan He, Wenhong Sun, Hao-Hsiung Lin, Shude Yao, Xiang Lu, Lingyu Wan and Zhechuan Feng
Investigation of HfO_2 Thin Films on Si by X-ray Photoelectron Spectroscopy, Rutherford Backscattering, Grazing Incidence X-ray Diffraction and Variable Angle Spectroscopic Ellipsometry
Reprinted from: *Crystals* **2018**, *8*, 248, doi:10.3390/cryst8060248 81

Hung-Pin Hsu, Der-Yuh Lin, Cheng-Ying Lu, Tsung-Shine Ko and Hone-Zern Chen
Effect of Lithium Doping on Microstructural and Optical Properties of ZnO Nanocrystalline Films Prepared by the Sol-Gel Method
Reprinted from: *Crystals* **2018**, *8*, 228, doi:10.3390/cryst8050228 . . . . **97**

Weijun Miao, Yiguo Li, Libin Jiang, Feng Wu, Hao Zhu, Hongbing Chen and Zongbao Wang
Epitaxial Crystallization of Precisely Methyl-Substituted Polyethylene Induced by Carbon Nanotubes and Graphene
Reprinted from: *Crystals* **2018**, *8*, 168, doi:10.3390/cryst8040168 . . . . **105**

Meng Ding, Zhen Guo, Lianqun Zhou, Xuan Fang, Lili Zhang, Leyong Zeng, Lina Xie and Hongbin Zhao
One-Dimensional Zinc Oxide Nanomaterials for Application in High-Performance Advanced Optoelectronic Devices
Reprinted from: *Crystals* **2018**, *8*, 223, doi:10.3390/cryst8050223 . . . . **119**

About the Special Issue Editor

Dong-Sing Wuu received his B.S., M.S., and Ph.D. degrees in Electrical Engineering from National Sun Yat-sen University, Taiwan, in 1985, 1987, and 1991, respectively. He has worked at Industry Technology Research Institute (ITRI, Taiwan), Da-Yeh University, National Formosa University, National Chung Hsing University (NCHU), and Nano and Advanced Materials Institute Limited (NAMI, Hong Kong), where he served in various positions. He is currently a lifetime distinguished Professor at Department of Materials Science and Engineering, NCHU. Dr. Wuu's steady advancements have been recognized in the form of numerous accolades, including the Distinguished Research Award (in the field of Photonics) in 2009, and the Excellent Technology Transfer Awards in 2007, 2010, and 2011 from the Ministry of Science & Technology (Taiwan). He also received the 2012 Annual Photonics Award from Taiwan Photonics Society, 2015 Annual National Teacher's Award from the Ministry of Education (Taiwan), and the 2017 Distinguished Professor Award from the Chinese Institute of Engineers. He has served as Fellow of the International Society for Optics and Photonics (SPIE), Optical Society (OSA), Institute of Physics (IOP), Institute of Engineering and Technology (IET), Thin Films Society (TFS), and Australian Institute of Energy (AIE). He has published over 300 SCI journal papers in addition to 6 book chapters, and presented 50 invited talks. He has been awarded over 150 international patents, of which 30 have been transferred to various companies. His current research interests include functional oxide material growth, thin-film technology, and solid-state optoelectronic devices.

Preface to "Functional Oxide Based Thin-Film Materials"

Functional oxide-based thin-film materials are extraordinary multifunctional crystals with a huge range of emerging application domains, including as sensors, displays, light emitters, photovoltaics, nanotechnology, spintronics, piezoelectric motors, biotechnology, capacitors, transparent electronics, and next-generation memories. These materials served as the original inspiration for the Special Issue "Functional Oxide-Based Thin-Film Materials". Functional oxide crystals have numerous favorable properties, including good transparency, high conductivity, wide bandgap, and strong luminescence. Thin-film oxide materials have been successfully grown on various substrates by hydrothermal, sol–gel, chemical bath deposition, sputtering, atomic layer deposition, pulsed laser deposition, chemical vapor deposition (CVD), hot filament CVD, plasma-enhanced CVD methods, among others. A number of breakthroughs over the recent years have driven an exponential increase in the research activity in this field. For this Special Issue "Functional Oxide-Based Thin-Film Materials", specialists working with device applications came together to shed light on the properties and behavior of thin-film oxides. This collections of papers covers many aspects of thin-film science and technology, from thin film to nanostructure and from material properties to optoelectronic applications, thus reflecting the numerous varied interests of the community of scientists active in this field.

Dong-Sing Wuu
Special Issue Editor

Editorial

Special Issue Editorial: Functional Oxide Based Thin-Film Materials

Dong-Sing Wuu

Department of Materials Science and Engineering, National Chung Hsing University, 145 Xingda Road, Taichung 40227, Taiwan; dsw@dragon.nchu.edu.tw

Received: 8 March 2020; Accepted: 10 March 2020; Published: 12 March 2020

Protective oxide coatings, such as Al_2O_3 and Y_2O_3 coatings, are widely used in semiconductor industries because of their hardness, high wear resistance, dielectric strength, high corrosion resistance, and chemical stability for plasma chambers [1]. Thin-film oxide barrier coatings with ultra-low permeation also received much attention in the industries of pharmaceuticals, food, beverage packaging, and organic light-emitting display applications [2]. Besides these protective functions, the oxide thin-film materials and nanostructures are extraordinary multifunctional crystals with a huge range of emerging application domains, such as sensors, displays, light emitters, photovoltaics, spintronics, piezoelectric motors, biotechnology, capacitors, transparent electronics, and next-generation memories [3]. These become the original motivation for proposing this Special Issue—Functional Oxide-Based Thin-Film Materials.

The functional oxide crystal has several favorable properties, including good transparency, high conductivity, a wide bandgap, and strong luminescence. Thin-film oxide materials have been grown on various substrates by hydrothermal, sol–gel, chemical bath deposition, sputtering, atomic layer deposition, pulsed laser deposition, chemical vapor deposition (CVD), hot filament CVD, plasma-enhanced CVD, etc. A number of breakthroughs over the past few years have driven an exponential energy in the research activity of this field.

This Special Issue on Functional Oxide-Based Thin-Film Materials aims at gathering together some of the specialists working with device applications, to shed light on the properties and behavior of thin-film oxides. The papers cover many aspects of thin-film science and technology from thin film to nanostructure, from material properties to optoelectronic applications, thus reflecting the many interests of the community of scientists active in the field.

Gallium oxide, a wide bandgap material, has received much attention in several possible applications. In [4], Lin and co-workers report the porous GaN layers surrounding the mesa region were transformed into insulating GaOx layers in a reflector structure through a lateral photoelectrochemical (PEC) oxidation process. The electroluminescence emission intensity was localized at the central mesa region by forming the insulating GaOx layers in a reflector structure as a current confinement aperture structure. The PEC light-emitting device structure with a current-confined aperture surrounded by insulating GaOx layers has the potential for InGaN resonance cavity light source applications. In [5], Wuu and co-workers report the synthesis and characterization of gallium oxide (Ga_2O_3) thin films fabricated on glass substrates using a combination of chemical bath deposition and a post-annealing process. The nanocrystalline α-Ga_2O_3 films obtained from GaOOH after annealing showed the photodegraded 90% of the methylene blue in a solution irradiated under an ultraviolet lamp for 5 h. These results suggest that α-Ga_2O_3 thin films can be very useful alternatives for the photocatalytic degradation of dyes during wastewater treatment.

Zinc oxide (ZnO) is a versatile and inexpensive semiconductor with a wide direct band gap that has applicability in several scientific and technological fields. In [6], Amakali and co-workers report the synthesis of ZnO thin films via two simple and low-cost synthesis routes, i.e., the molecular precursor method and the sol–gel method, which were deposited successfully on microscope glass substrates.

In [7], Díaz-Becerril et al. have investigated the tantalum-doped ZnO structures synthesized on silicon substrates by using a hot filament chemical vapor deposition reactor. Green photoluminescence (PL) was observed by the naked eye when Ta-doped samples were illuminated by ultraviolet radiation and confirmed by photoluminescence spectra. The PL intensity on the Ta-doped ZnO increased from those undoped samples up to eight times and the resistivity and the sheet resistance also decrease when there is a greater amount of tantalum in the film.

An interesting investigation into the series of nanometer scale (33–70 nm) HfO_2 thin films grown on Si substrates under different conditions by atomic vapor deposition (AVD) is reported by Feng et al. in [8]. The comprehensive studies demonstrate that appropriate substrate temperature and oxygen flow are essential to the structure, chemical composition, and optical constants from the surface and interface of the HfO_2 films deposited by AVD. This work with integrated experiment measurements and analyses has enhanced our understanding of AVD-grown HfO_2 advanced materials.

An entirely different perspective on thin film oxides is presented by Chang and Chan. In their paper, [9], they investigated the heavily doped wide band gap semiconductors like aluminum-doped zinc oxide (AZO) and tin-doped indium oxide (ITO) for low emissivity glass (low-e glass), which is often used in energy-saving buildings. The emissivity and average visible transmittance for H_2/N_2 = 100/100 plasma annealed AZO/ITO were 0.07% and 80%, respectively, lying in the range of commercially used low emissivity glass.

Supercapacitors, excellent energy storage devices, can effectively alleviate the current energy crisis. Based on their obvious advantages, such as simple design, high-power density, a long cycling lifetime, and a short charge/discharge rate, supercapacitors have attracted much research interest in recent years. Zheng and co-workers in [10] have synthesized a NiO@ZnO (NZO) hybrid with different reaction times by a green hydrothermal method. A highest energy density of 27.13 Wh kg^{-1} can be achieved at a power density of 321.42 W kg^{-1}. These desirable electrochemical properties demonstrate that the NZO hybrid is a promising electrode material for a supercapacitor.

Finally, we also presented a contribution concerned with a review article, using 1D zinc oxide (ZnO) as a representative nanomaterial [11]. This article reviews the fabrication methods of 1D ZnO nanostructures—including chemical vapor deposition, hydrothermal reaction, and electrochemical deposition—and the influence of the growth parameters on the growth rate and morphology. Important applications of 1D ZnO nanostructures in optoelectronic devices are described. Several approaches to improve the performance of 1D ZnO-based devices, including surface passivation, localized surface plasmons, and the piezo-phototronic effect, are summarized.

To conclude, I believe that this Special Issue on Functional Oxide-Based Thin-Film Materials touches on the latest advancements in several aspects related to material science: the synthesis of novel oxide, photoluminescence characteristics, photocatalytic ability, energy storage, light emitter studies, low emissivity glass coatings, and investigations of both nanostructure and thin film properties. I wish to express my deepest and sincere gratitude to all authors who contributed, for having submitted manuscripts of such excellent quality. I also wish to thank the Editorial Office of *Crystals* for the fast and professional handling of the manuscripts during the whole submission process and for the help provided.

Conflicts of Interest: The authors declare no conflict of interest

References

1. Lin, T.-K.; Wang, W.-K.; Huang, S.-Y.; Tasi, C.-T.; Wuu, D.-S. Comparison of Erosion Behavior and Particle Contamination in Mass-Production CF4/O2 Plasma Chambers Using Y2O3 and YF3 Protective Coatings. *Nanomaterials* **2017**, *7*, 183. [CrossRef] [PubMed]
2. Hsu, C.-H.; Lin, Y.-S.; Wu, H.-Y.; Zhang, X.-Y.; Wu, W.-Y.; Lien, S.-Y.; Wuu, D.-S.; Jiang, Y.-L. Deposition of Silicon-Based Stacked Layers for Flexible Encapsulation of Organic Light Emitting Diodes. *Nanomaterials* **2019**, *9*, 1053. [CrossRef] [PubMed]
3. Peng, W.-C.; Chen, Y.-C.; He, J.-L.; Ou, S.-L.; Horng, R.-H.; Wuu, D.-S. Tunability of p- and n-channel TiOx Thin Film Transistors. *Sci. Rep.* **2018**, *8*, 9255. [CrossRef] [PubMed]
4. Chen, Y.-Y.; Jhang, Y.-C.; Wu, C.-J.; Chen, H.; Lin, Y.-S.; Lin, C.-F. Fabrication of GaOx Confinement Structure for InGaN Light Emitter Applications. *Crystals* **2018**, *8*, 418. [CrossRef]
5. Yeh, C.-Y.; Zhao, Y.-M.; Li, H.; Yu, F.-P.; Zhang, S.; Wuu, D.-S. Growth and Photocatalytic Properties of Gallium Oxide Films Using Chemical Bath Deposition. *Crystals* **2019**, *9*, 564. [CrossRef]
6. Amakali, T.; Daniel, L.S.; Uahengo, V.; Dzade, N.Y.; Leeuw, N.H. Structural and Optical Properties of ZnO Thin Films Prepared by Molecular Precursor and Sol–Gel Methods. *Crystals* **2020**, *10*, 132. [CrossRef]
7. Herrera, V.; Díaz-Becerril, T.; Reyes-Cervantes, E.; García-Salgado, G.; Galeazzi, R.; Morales, C.; Rosendo, R.; Nieto-Caballero, F.G. Highly Visible Photoluminescence from Ta-Doped Structures of ZnO Films Grown by HFCVD. *Crystals* **2018**, *8*, 395. [CrossRef]
8. Luo, X.; Li, Y.; Yang, H.; Liang, Y.; He, K.; Sun, W.; Lin, H.-H.; Yao, S.; Lu, X.; Wan, L.; et al. Investigation of HfO2 Thin Films on Si by X-ray Photoelectron Spectroscopy, Rutherford Backscattering, Grazing Incidence X-ray Diffraction and Variable Angle Spectroscopic Ellipsometry. *Crystals* **2018**, *8*, 248. [CrossRef]
9. Chang, S.-C.; Chan, H.-T. Effect of Nitrogen Flow in Hydrogen/Nitrogen Plasma Annealing on Aluminum-Doped Zinc Oxide/Tin-Doped Indium Oxide Bilayer Films Applied in Low Emissivity Glass. *Crystals* **2019**, *9*, 310. [CrossRef]
10. Zheng, J.; Zhang, R.; Cheng, K.; Xu, Z.; Yu, P.; Wang, X.; Niu, S. Research on the High-Performance Electrochemical Energy Storage of a NiO@ZnO (NZO) Hybrid Based on Growth Time. *Crystals* **2019**, *9*, 47. [CrossRef]
11. Ding, M.; Guo, Z.; Zhou, L.; Fang, X.; Zhang, L.; Zeng, L.; Xie, L.; Zhao, H. One-Dimensional Zinc Oxide Nanomaterials for Application in High-Performance Advanced Optoelectronic Devices. *Crystals* **2018**, *8*, 223. [CrossRef]

© 2020 by the author. Licensee MDPI, Basel, Switzerland. This article is an open access article distributed under the terms and conditions of the Creative Commons Attribution (CC BY) license (http://creativecommons.org/licenses/by/4.0/).

Article

Structural and Optical Properties of ZnO Thin Films Prepared by Molecular Precursor and Sol–Gel Methods

Theopolina Amakali [1,*], Likius. S. Daniel [1], Veikko Uahengo [1], Nelson Y. Dzade [2,*] and Nora H. de Leeuw [2,3,*]

1. Department of Chemistry and Biochemistry, University of Namibia, 340 Mandume Ndemufayo Avenue, Windhoek 9000, Namibia; daniels@unam.na (L.S.D.); vuahengo@gmail.com (V.U.)
2. School of Chemistry, Cardiff University, Main Building, Park Place, Cardiff CF10 3AT, UK
3. Department of Earth Sciences, Utrecht University, Princetonplein 8A, 3584 CB Utrecht, The Netherlands
* Correspondence: Inaandtuli@gmail.com (T.A.); DzadeNY@cardiff.ac.uk (N.Y.D.); DeLeeuwN@cardiff.ac.uk (N.H.d.L.)

Received: 12 November 2019; Accepted: 18 February 2020; Published: 22 February 2020

Abstract: Zinc oxide (ZnO) is a versatile and inexpensive semiconductor with a wide direct band gap that has applicability in several scientific and technological fields. In this work, we report the synthesis of ZnO thin films via two simple and low-cost synthesis routes, i.e., the molecular precursor method (MPM) and the sol–gel method, which were deposited successfully on microscope glass substrates. The films were characterized for their structural and optical properties. X-ray diffraction (XRD) characterization showed that the ZnO films were highly *c*-axis (0 0 2) oriented, which is of interest for piezoelectric applications. The surface roughness derived from atomic force microscopy (AFM) analysis indicates that films prepared via MPM were relatively rough with an average roughness (Ra) of 2.73 nm compared to those prepared via the sol–gel method (Ra = 1.55 nm). Thin films prepared via MPM were more transparent than those prepared via the sol–gel method. The optical band gap of ZnO thin films obtained via the sol–gel method was 3.25 eV, which falls within the range found by other authors. However, there was a broadening of the optical band gap (3.75 eV) in thin films derived from MPM.

Keywords: zinc oxide; molecular precursor method; crystallite size; optical band gap

1. Introduction

Zinc oxide (ZnO) is one of the most widely researched semiconductor oxides, owing to its many versatile and attractive properties, such as high chemical and thermal stability, non-toxicity [1], ease of preparation, tunable direct wide band gap (3.4 eV), and high transparency in the visible region [2]. These properties saw ZnO thin films fabricated for various industrial applications, e.g., in optoelectronic devices [3], lasers, gas sensors, and ultraviolet (UV) light emitters [4], and as protective surface coatings [5]. ZnO plays important roles in a number of solar cell systems, such as silicon-based solar cells (first-generation), thin films (second generation), and organic multi-junction, dye-sensitized (third-generation) systems, either as a transparent conductive oxide (TCO) or as a junction for exciton separation [6].

Several techniques were proposed for the fabrication of ZnO thin films to modify its optical properties, including both physical and chemical processes. Physical methods include sputtering techniques [7–9], pulsed laser deposition [10], and molecular beam epitaxy (MBE) [11]. However, these methods are often complex and require costly vacuum equipment. In contrast, ZnO thin films can also be prepared via chemical methods, such as chemical bath deposition [12], chemical vapor deposition [13], atomic layer deposition [14], spray pyrolysis [15], printing [16], sol–gel spin coating [17], and electrochemical deposition [18], which are simpler methods and generally less costly. The conventional sol–gel method is often preferred over other chemical methods, due to its simplicity, lower crystallization temperature, and compositional control [19].

Structural and optical properties of ZnO thin films prepared via the sol–gel technique, using a variety of inorganic and organic precursors under different deposition conditions, were reported in the literature [20–22]. Post-heat treatment is one of the factors that plays a crucial role in the structural and optical properties of the ZnO thin films through the modification of their crystallinity and surface roughness. Thermal gravimetric and differential thermal gravimetric analysis (TGA/DTA) of ZnO gelatum by Meng et al. [23] showed that annealing the gel at 500 °C ensured complete removal of all the hydroxyl bonds and residues of organic matter. Chaitra et al. [24] investigated the role of annealing temperature (300, 400, and 500 °C) on the structural and optical properties of sol–gel-derived ZnO thin films. Thin films annealed at 500 °C showed tensile stress and a preferential orientation toward the (0 0 2) plane along the c- axis. Raoufi and Raoufi [25], using the same heat treatment conditions, obtained dense ZnO thin films with a (0 0 2) preferred orientation that was enhanced at 500 °C. Other authors reported similar observations [26,27]. In these studies, the authors reported that an increase in the annealing temperature reduced the full width at half-maximum (FWHM), increased grain size and crystallite size, and, as a result, increased the surface roughness. An increase in annealing temperature also decreased the optical band gap. However, cracking of the thin films during the annealing step remains a challenge, whereas the presence of interfaces within the thin films during drying can cause a reduction in the optical transparency [28]. It is, thus, important to explore other facile fabrication methods to overcome some of these challenges, and we, therefore, explored the fabrication of ZnO thin films using the molecular precursor method (MPM), which is a wet chemical process for the preparation of metal oxide thin films [29,30]. MPM, unlike the sol–gel method, is based on the use of metal complexes in coating solutions that are known to have excellent stability, homogeneity, and miscibility, owing to the metal complex anions which have better stability and can be dissolved in volatile solvents by combining them with the appropriate alkylamines [31].

The molecular precursor method (MPM) was employed successfully in the fabrication of copper oxides [31], as well as homogeneous and crack-free titania thin films [32]. Mashiyama and co-workers [33] reported the successful fabrication via MPM of Mg–Zn–O thin films, which exhibited good properties for applications as near-infrared UV-transparent electrodes for GaN-based UV-light-emitting diodes (LEDs). In 2012, Taka and co-authors [34] reported the use of MPM in the fabrication of composite ZnO thin films with dispersed Ag for application in GaInN blue LEDs. However, to the authors' knowledge, to date, there are no reports on the fabrication and optical properties of pure ZnO thin films using the molecular precursor method.

In the present work, we report the growth of ZnO thin films on microscope slide substrates fabricated via the sol–gel method and MPM, as well as the investigation of their structural properties, surface morphology, and optical properties through X-ray diffraction (XRD), atomic force microscopy (AFM), and UV/visible light (Vis) spectra. Depending on the observed properties, the thin films thus fabricated could find application as sensors and catalysts, or in solar cells.

2. Materials and Synthesis Methods

2.1. Materials

Ethylenediaminetetraacetic acid (EDTA), zinc acetate dihydrate $Zn(CH_3COO)_2 \cdot 2H_2O$, 2-methoxyethanol, monoethalamine (MEA), absolute ethanol, and methanol were all purchased from Merck, Darmstadt, Germany. All reagents were of analytical grade and used without further purification. Microscope glass slides (26 mm × 38 mm) from B&C, Germany, were used as substrates for the deposition of ZnO thin films.

2.2. Synthesis of ZnO Thin Films

2.2.1. Preparation of Precursor Solution for Fabrication of ZnO Thin Film Using Sol–Gel Method

The Zn precursor solution (0.4 M) was prepared according to the procedure of Khan et al. [22] with modifications. The procedure used zinc acetate dihydrate, a 1:1 ethanol–methanol mixture, and dibutylamine as the Zn source, solvent, and stabilizer, respectively. The weight ratio of zinc acetate dihydrate to dibutylamine was fixed at 1.0. The solution was then stirred for 2 h at 50 °C and aged for 24 h at room temperature. This procedure yielded a light-yellow homogeneous viscous gel.

2.2.2. Preparation of Precursor Solution for Fabrication of ZnO Thin Films Using MPM

The procedure was based on the method of Sato et al. [29] with modifications. A Zn precursor solution (0.4 M) was prepared using EDTA, zinc acetate dihydrate, 2-methoxyethanol, and monoethanolamine as a complexing agent, Zn source, solvent, and stabilizer, respectively. The mole ratio of zinc acetate dihydrate to EDTA was 1:1, while it was 1:2.45 for monoethanolamine. In this procedure, EDTA was dissolved in 2-methoxyethanol and monoethanolamine was added; then, the solution was refluxed with constant stirring at 55 °C for 30 min. The solution was cooled to room temperature and zinc acetate dihydrate was added. The solution was refluxed further at 65 °C for 2 h, and a clear orange solution was obtained.

2.3. Film Fabrication by Coating and Heat Treatment

Firstly, 200 µL of the precursor solutions obtained from the MPM and the sol–gel routes were spin-coated onto pre-cleaned glass microscope slides using a double step mode, first at 500 rpm for 5 s and then at 2500 rpm for 30 s. The films were dried in a preheated oven at 150 °C for 10 min. The coating and drying steps were repeated up to 10 times to improve the thickness of the films. The thin films were then annealed at 500 °C for 1 h. Earlier studies demonstrated that annealing temperatures of around 500 °C generally result in the formation of well-crystallized metal-oxide thin films [20–22].

2.4. Characterization of ZnO Thin Films

The crystal structure of the fabricated ZnO thin films was investigated using a Bruker D8 Advance X-ray diffractometer (XRD) with CuKα radiation λ = 1.5402 Å. The surface morphology and topography were evaluated with a Bruker Dimension Edge atomic force microscope (AFM) with ScanAsyst TM and the Hitachi S4800 FE-SEM. The optical transmittance was measured on a Perkin-Elmer Lambda 750 UV-Vis/near-infrared (NIR) spectrophotometer in the range of 300 nm to 700 nm.

3. Results and Discussion

3.1. Crystal Structure and Particle Size

Figure 1 shows the XRD patterns of ZnO thin films on the microscope glass prepared via the MPM and sol–gel methods. The XRD spectra of sol–gel-derived thin films indicate the presence of three dominant peaks, corresponding to diffraction planes of (1 0 0), (0 0 2), and (1 0 1), showing the growth of ZnO crystallites along different directions. The peaks correspond to those of the standard ZnO (JCPDS 36-1451), and the typical hexagonal wurtzite structure was, therefore, inferred for the thin films from the XRD patterns. In this study, the baseline was not horizontal due to the amorphous glass substrate used. O'Brien et al. [26] observed a similar broad feature between 20° and 40°, which was attributed to the amorphous nature of the glass substrate used. Only the (0 0 2) peak was observed for the MPM-derived thin films, which is the kinetically favored orientation along the c-axis [35].

It was reported that, in ZnO, unlike thin films synthesized using short-chain alcohols such as ethanol, films synthesized using solvents with high boiling points (e.g., 2-methoxyethanol) show a very strong preferential orientation along the (0 0 2) plane [36–38]. This finding was attributed to the slow evaporation of high-molecular-weight alcohol upon heating, allowing structural orientation of the film before crystallization. In addition, this preferential orientation observed for ZnO grown on an amorphous glass substrate is due to the presence of non-bridging oxygen atoms in the glass substrate which assist the growth of ZnO along the (0 0 2) plane [39]. The (0 0 2) diffraction peak was widely observed as a preferred orientation in solution-grown ZnO [40,41], and the appearance of the (0 0 2) diffraction peak suggests that the surface free energy of this plane is the lowest both in the sol–gel- and in the MPM-fabricated ZnO thin films, thereby resulting in its preferred orientation.

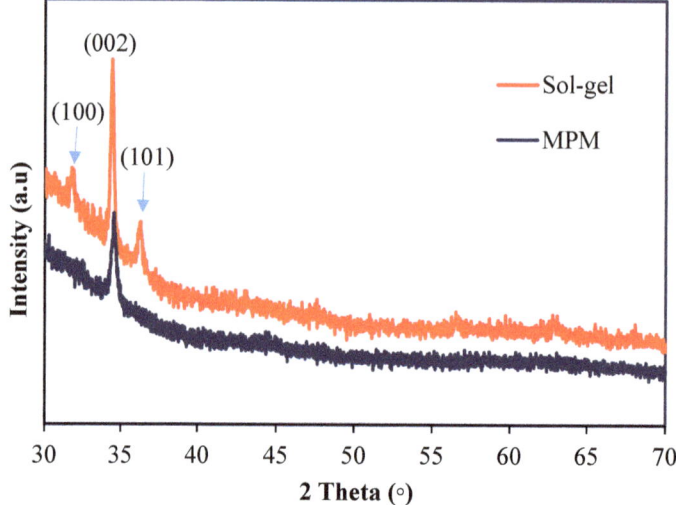

Figure 1. X-ray diffraction (XRD) patterns of ZnO fabricated via molecular precursor method (MPM) and sol–gel method.

The crystallite size (D) was estimated from the Debye–Scherrer equation.

$$D = \frac{0.9\lambda}{\beta cos\theta} \quad (1)$$

where D is the crystallite size, λ (1.5418 nm) is the wavelength of the incident X-ray beam CuK_a, β is the full width at half-maximum, and θ is the Bragg diffraction angle. The evaluated structural parameters of the thin films synthesized via the two methods are presented in Table 1. The line broadening at half the maximum intensity (FWHM) was relatively large (4.535 $\times 10^{-3}$ rad) in thin films obtained from the MPM compared to those obtained from the sol–gel method (3.664 $\times 10^{-3}$ rad).

Table 1. Structural parameters of ZnO thin films prepared via two different synthesis routes. FWHM—full width at half-maximum.

Parameter	MPM	Sol–Gel
2θ (°)	34.66	34.52
FWHM β ($\times 10^{-3}$ rad)	4.535	3.664
Crystallite size D (nm)	32.00	39.60
Dislocation density δ ($\times 10^{14}$ lines/m^2)	9.77	6.38
Strain ε ($\times 10^{-3}$)	1.89	1.53

This indicates that crystallinity was better in the sol–gel-derived thin films, as the FWHM is inversely related to the crystallite size (Equation (1)). The crystallite size for the thin films synthesized through the MPM method was significantly smaller (32 nm) compared to that obtained through the sol–gel method (39.6 nm). A larger crystallite size of the sol–gel-derived ZnO thin films is an indication of less ordered grains and fewer micro-defects in the grains [39], which translates into a small value for the dislocation density and strain given in Table 1.

Figure 2 shows the scanning electron microscopy (SEM) image of a ZnO thin film synthesized via the molecular precursor method. The surface appears to be microporous with inhomogeneous spherical grains of different sizes. The average value of the crystallite size value obtained from the scanning electron microscopy (SEM) analysis was about 34 nm, which is quite close to that obtained from the XRD analysis (32 nm). Differences in thermal expansion represent one of the factors that affect the bonding between the coating and the underlying micro glass substrate. The influence of the difference in the thermal expansion coefficient between the ZnO film and the substrate is negligible, because of the very thin nature of the deposited ZnO thin film with a height of 7.67 nm (Figure 3). As a result, there is stable bonding between the ZnO thin film and the micro glass substrate. The exact bonding mechanism of the thin films and supportive substrate in the molecular precursor method is not clear. Further studies related to the observation of the interface between the coated ZnO thin film and glass substrate are needed.

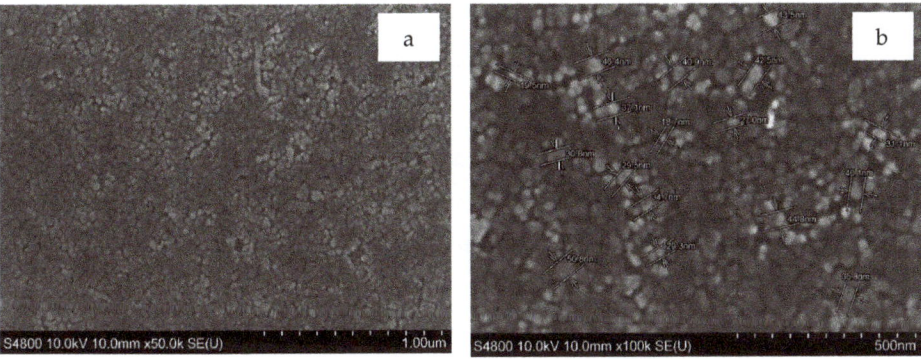

Figure 2. SEM image of (**a**) microporous ZnO thin film; (**b**) different grain sizes fabricated via MPM method.

Consistent with our results, previous reports showed that the molecular precursor method produces thin films of small crystallite size due to a rapid nucleation process [42]. The main difference between the sol–gel and the molecular precursor method is the absence of polymerization in the MPM solution, ensuring that the precursor film formed after coating remains amorphous and the solution remains stable over a long time [43]. The shrinkage rate and the packing of the precursor film on the substrate depend on the chain length of the alkylamine used in the synthesis [43]. It was suggested that the crystallite size of the metal oxide obtained via the MPM method will be smaller than that obtained from the sol–gel process [44], because the nucleation process in the sol–gel method during heat treatment is slower, as the polymeric chains rearrange to form the basic core structure of the metal oxide, resulting in relatively large crystallites.

The dislocation density (δ) and lattice strain (ε) in the thin films were estimated using the following equations:

$$\delta = \frac{1}{D^2} \qquad (2)$$

$$\varepsilon = \frac{\beta \cos\theta}{4} \qquad (3)$$

As shown in Table 1, ZnO thin films obtained from the sol–gel method showed smaller values for the dislocation density and lattice strain than those obtained from the molecular precursor method. The dislocation density and lattice strain represent flaws and levels of defects in the crystal structure of the thin films, with smaller values indicating better crystallinity of the thin films [41]. Thus, it can be inferred that the thin films obtained from the sol–gel method are of better quality than those obtained from the molecular precursor method, potentially because the formation of crystallites of larger sizes allows the release of strain energy and relaxation between grains in the thin films [42].

Topographical features of the ZnO thin films were measured with AFM in tapping mode over a scan area of 1 µm². Figure 3 shows the micrographs of the thin films prepared via the MPM and sol–gel methods. Based on the surface roughness and height recorded in Table 2, it is evident that the thin films obtained from MPM have a high surface roughness, with the root mean square (RMS) of the surface = 3.47 nm, compared to that obtained from the sol–gel method where RMS = 2.02 nm.

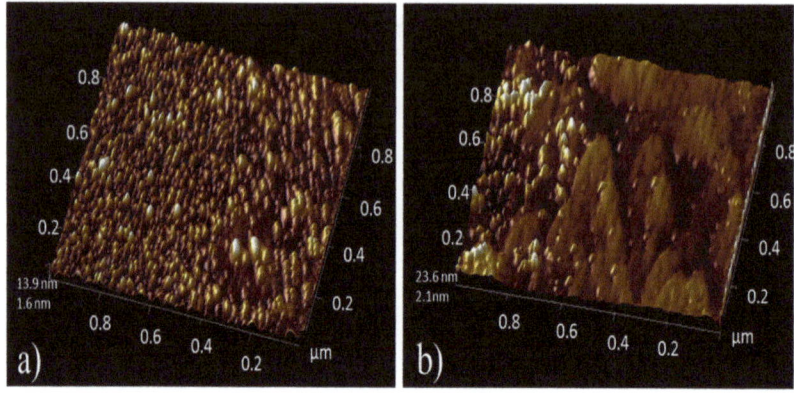

Figure 3. Atomic force microscopy (AFM) image (1 µm × 1 µm) of ZnO thin films prepared via (**a**) MPM and (**b**) sol–gel techniques.

Table 2. Surface roughness of MPM- and sol–gel-derived ZnO thin films.

Roughness Profile	MPM	Sol–Gel
Root mean square, RMS (nm)	3.47	2.02
Roughness average, Ra (nm)	2.73	1.50
Height (nm)	7.67	7.69

An inverse relationship between crystallite size and surface roughness was observed in this study, with the same findings reported by other authors [41]. It is, however, worth noting that the surface roughness of the ZnO thin film may also be affected by other factors and does not only depend on the crystallite size. Other parameters such as shape of the nanostructures, deposition conditions, and chemical reactions in the precursor solutions may also affect the surface roughness of ZnO thin films [43,44]. The influence of complexing agents on the grain size and surface roughness of ZnO thin films was described in detail by Nesheva et al. [39].

3.2. Optical Properties

Figure 4 shows the optical transmittance of the ZnO thin films synthesized via MPM and the sol–gel method. All transmittance spectra showed intensive absorption in the UV region, as expected for ZnO. The average transmittance for the ZnO thin films was calculated in the visible wavelength range of 400–700 nm. All thin films were highly transparent with an average transmittance of about 88% for the sol–gel-derived thin films and 90% for the MPM-derived thin films. Both films are, thus, suitable for optoelectronic devices, e.g., as window layers in solar cells. Salam et al. [45] determined the absorption edge of sol–gel-synthesized intrinsic (i-ZnO) and aluminum-doped zinc oxide (Al:ZnO) thin films mined to be at ~370 nm with ≥80% transmittance in the visible and near-infrared regions of the spectrum. Nearly 100% transmittance across the visible range was reported for ZnO thin films synthesized from a one-step spin-coating pyrolysis technique using zinc neodecanoate precursor [46]. A high transmittance (~92%) of ZnO nanowire arrays on indium tin oxide (ITO) substrate in the visible region was reported by Chen et al. [47]. The effect of annealing temperature on the structural, morphological, and optical properties of spray-pyrolized Al-doped ZnO thin films was systematically investigated by Kabir et al. [48]. It was shown that the annealing temperature affects the optical transmittance of the Al-doped ZnO thin films with the average transmittance value within the visible region found to be 70.3% for the as-deposited film [48].

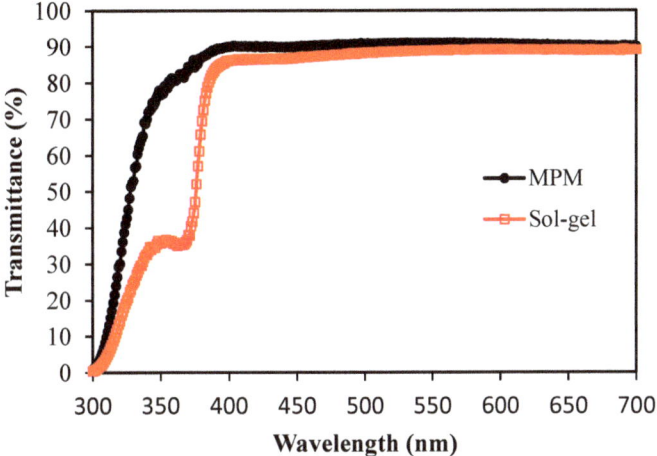

Figure 4. Transmittance spectra of ZnO fabricated via MPM and sol–gel methods.

The optical band gap energy (E_g) of the ZnO was determined from Tauc's equation (Equation (4)) for direct band gap semiconductors.

$$Ah\nu = A(H\nu - Eg)^{\frac{1}{2}}, \qquad (4)$$

where α is the absorption coefficient, $h\nu$ is the incident photon energy, and A is a constant.

The optical band gap was then estimated by plotting $(\alpha h\nu)^2$ against the photon energy ($h\nu$) and extrapolating the linear portion of the curve to the photon energy axis, as shown in Figure 5. The presence of a single slope in the plot for ZnO fabricated using MPM suggests that, unlike in the sol–gel method, films in MPM have direct and allowed transitions. The band gap energy (eV) is obtained by extrapolating the straight-line portion of the plot to the zero absorption coefficient. Thus, the band gap values of the ZnO thin films fabricated via the sol–gel method and MPM were found to be 3.25 and 3.75 eV, respectively. However, the Tauc plot for thin films obtained from the sol–gel method shows a second linear segment, which could be due to sub-band gaps from a possible secondary phase [49]. The value of the optical band gap for the ZnO fabricated via the sol–gel method agrees with what is reported in the literature [50,51]. ZnO thin films that can absorb in the visible range of the spectrum make these materials suitable as a window layer in solar cells. However, the band gap for the ZnO thin film fabricated using MPM was a bit higher. There are two possible reasons for a large band gap value of the film: (i) owing to an axial strain effect from lattice deformation as suggested for ZnO films [52], or (ii) owing to a change in the density of semiconductor carriers. Which of these scenarios is the most likely, however, requires further investigation. Large optical band gap values (3.52–3.71 eV) were also reported by Akhtar et al. [53] and Wahab et al. [54]. Both groups attributed the blue shift of the band gap to residual strain defect and grain size confinement, although the grain sizes ranged between 28 nm and 150 nm. Samanta et al. [55] reported that the confinement regime of ZnO is much stronger when the particle size is much closer to the Bohr radius (2.3 nm). Hence, band gap broadening of ZnO thin films cannot be due to a quantum confinement effect, because the size of the ZnO crystallites is outside the quantum confinement regime, which is usually from 2 nm to 10 nm.

Given that the thin films in this study were very thin, the blue shift in the band gap could be due to the presence of interference from the amorphous glass substrate used. It is also possible that there could be a reformation of intermediate phases or new amorphous ZnO-like structures, as reported by Gonzalenz [56] and Nishio [37]. Nonetheless, other factors, such as variations in shape and size, may also play a role [57,58].

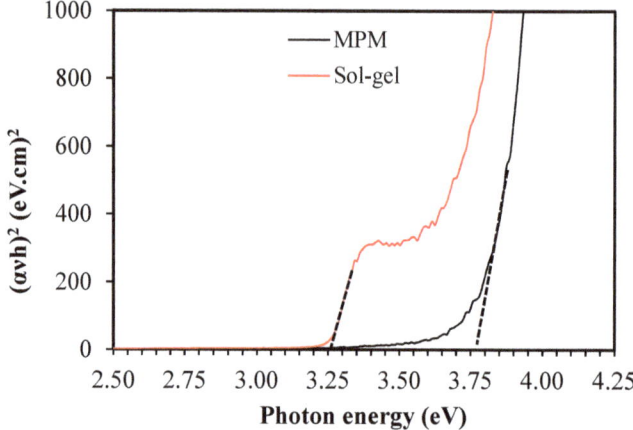

Figure 5. Plots of $[\alpha vh]^2$ versus photon energy of ZnO thin films fabricated via MPM and sol–gel methods.

4. Summary and Conclusions

ZnO thin films with a preferential (0 0 2) orientation along the c-axis deposited on microscope glass slides were synthesized via two synthesis routes: sol–gel and molecular precursor methods. It was demonstrated, based on the estimated structural parameters, such as crystallite size, dislocation density, strain, and surface roughness, that the sol–gel-derived ZnO thin films were of better quality and crystallinity than those prepared via the MPM method. Results from optical studies showed a larger band gap (3.75 eV) for thin films fabricated using MPM. Factors such as the synthesis and deposition conditions employed in this study could be responsible for the observed larger band gap. The observed high transparency of the as-prepared ZnO thin films (88% for sol–gel and 90% for MPM) in the visible region makes them suitable for use as transparent windows for various applications. Future investigations will characterize other optical properties such as the absorption coefficient and refractive index, as well as explore transition-metal doping to narrow the band gaps and to extend the absorption in the visible region.

Author Contributions: T.A. performed the experimental synthesis, characterization, and data analysis and wrote the paper. L.S.D., V.U., and N.Y.D. contributed to the study design and scientific discussions of the results. N.H.D.L. led the overall research program. All co-authors contributed to the manuscript. All authors have read and agreed to the published version of the manuscript.

Funding: The authors acknowledge the Royal Society and the UK Department for International Development, for a research grant under the Africa Capacity Building Initiative (ACBI), which funded this research.

Acknowledgments: This work made use of XRD and AFM facilities of the Botswana International University of Science and Technology (BIUST) and we are grateful to Foster Mbaiwa for his help with sample characterization. Sachin Rondiya is thanked for his assistance with the analyses of the XRD and AFM data. The SEM instrument used is based at Swansea University and Daniel R. Jones is appreciated for his help with SEM measurements.

Conflicts of Interest: The authors declare no conflicts of interest.

References

1. Yin, H.; Coleman, V.; Casey, P.; Angel, B.M.; Catchpoole, H.J.; Waddington, L.; McCall, M. A comparative study of the physical and chemical properties of nano-sized ZnO particles from multiple batches of three commercial products. *J. Nanoparticle Res.* **2015**, *17*, 96. [CrossRef]
2. Joshi, K.; Rawat, M.; Gautam, S.K.; Singh, R.; Ramola, R.; Mahajan, A. Band gap widening and narrowing in Cu-doped ZnO thin films. *J. Alloy. Compd.* **2016**, *680*, 252–258. [CrossRef]
3. Malik, G.; Mourya, S.; Jaiswal, J.; Chandra, R. Effect of annealing parameters on optoelectronic properties of highly ordered ZnO thin films. *Mater. Sci. Semicond. Process.* **2019**, *100*, 200–213. [CrossRef]
4. Rong, P.; Ren, S.; Yu, Q. Fabrications and Applications of ZnO Nanomaterials in Flexible Functional Devices-A Review. *Crit. Rev. Anal. Chem.* **2018**, *49*, 336–349. [CrossRef]
5. Ennaceri, H.; Erfurt, D.; Wang, L.; Köhler, T.; Taleb, A.; Khaldoun, A.; El Kenz, A.; Benyoussef, A.; Ennaoui, A.; Ahmed, D.E. Deposition of multifunctional TiO$_2$ and ZnO top-protective coatings for CSP application. *Surf. Coatings Technol.* **2016**, *298*, 103–113. [CrossRef]
6. Steiger, P.; Zhang, J.; Harrabi, K.; Hussein, I.; Downing, J.; McLachlan, M.A. Hydrothermally grown ZnO electrodes for improved organic photovoltaic devices. *Thin Solid Films* **2018**, *645*, 417–423. [CrossRef]
7. Mahdhi, H.; Djessas, K.; Ben Ayadi, Z. Synthesis and characteristics of Ca-doped ZnO thin films by rf magnetron sputtering at low temperature. *Mater. Lett.* **2018**, *214*, 10–14. [CrossRef]
8. Look, D.C. Mobility vs thickness in n + -ZnO films: Effects of substrates and buffer layers. *Mater. Sci. Semicond. Process.* **2017**, *69*, 2–8. [CrossRef]
9. Husna, J.; Aliyu, M.M.; Islam, M.A.; Chelvanathan, P.; Hamzah, N.R.; Hossain, M.S.; Karim, M.; Amin, N. Influence of annealing temperature on the properties of ZnO thin films grown by sputtering. *Energy Procedia* **2012**, *25*, 55–61. [CrossRef]
10. Labis, J.; Hezam, M.; Al-Anazi, A.; Al-Brithen, H.; Ansari, A.A.; El-Toni, A.M.; Enriquez, R.; Jacopin, G.; Alhoshan, M. Pulsed laser deposition growth of 3D ZnO nanowall network in nest-like structures by two-step approach. *Sol. Energy Mater. Sol. Cells* **2015**, *143*, 539–545. [CrossRef]
11. Opel, M.; Geprägs, S.; Althammer, M.; Brenninger, T.; Gross, R. Laser molecular beam epitaxy of ZnO thin films and heterostructures. *J. Phys. D Appl. Phys.* **2013**, *47*, 34002. [CrossRef]
12. Ortega-López, M.; Morales-Acevedo, A. Properties of ZnO Thin Films for Solar Cells Grown by Chemical Bath Deposition. In Proceedings of the Conference Record of the Twenty Sixth IEEE Photovoltaic Specialists Conference, Anaheim, CA, USA, 29 September–3 October 1997; pp. 555–558.
13. Wu, B.; Zhuang, S.; Chi, C.; Shi, Z.; Jiang, J.-Y.; Dong, X.; Li, W.-C.; Zhang, Y.; Zhang, B.-L.; Du, G.-T. The growth of ZnO on stainless steel foils by MOCVD and its application in light emitting devices. *Phys. Chem. Chem. Phys.* **2016**, *18*, 5614–5621. [CrossRef] [PubMed]
14. Graniel, O.; Fedorenko, V.; Viter, R.; Iatsunskyi, I.; Nowaczyk, G.; Weber, M.; Załęski, K.; Jurga, S.; Smyntyna, V.; Miele, P.; et al. Optical properties of ZnO deposited by atomic layer deposition (ALD) on Si nanowires. *Mater. Sci. Eng. B* **2018**, *236*, 139–146. [CrossRef]
15. Bedia, A.; Bedia, F.; Aillerie, M.; Maloufi, N.; Benyoucef, B. Morphological and optical properties of ZnO thin films prepared by spray pyrolysis on glass substrates at various temperatures for integration in solar cell. *Energy Procedia* **2015**, *74*, 529–538. [CrossRef]
16. Ismail, B.; Abaab, M.; Rezig, B. Structural and electrical properties of ZnO films prepared by screen printing technique. *Thin Solid Films* **2001**, *383*, 92–94. [CrossRef]
17. Liu, A.; Zhang, J.; Wang, Q. Structural and optical properties of zno thin films prepared by different sol-gel processes. *Chem. Eng. Commun.* **2010**, *198*, 494–503. [CrossRef]
18. Lei, J.F.; Wang, Z.W.; Li, W.S. Controlled fabrication of ordered structure-based ZnO films by electrochemical deposition. *Mater. Sci. Semicond. Process.* **2014**, *573*, 74–78.
19. Znaidi, L. Sol–gel-deposited ZnO thin films: A review. *Mater. Sci. Eng. B* **2010**, *174*, 18–30. [CrossRef]
20. Bahadur, H.; Srivastava, A.K.; Sharma, R.K.; Chandra, S. Morphologies of sol–gel derived thin films of ZnO using different precursor materials and their nanostructures. *Nanoscale Res. Lett.* **2007**, *2*, 469–475. [CrossRef]
21. Li, H.; Wang, J.; Liu, H.; Zhang, H.; Li, X. Zinc oxide films prepared by sol–gel method. *J. Cryst. Growth* **2005**, *275*, e943–e946. [CrossRef]

22. Khan, Z.R.; Khan, M.S.; Zulfequar, M.; Khan, M.S. Optical and structural properties of ZnO thin films Fabricated by Sol-Gel Method. *Mater. Sci. Appl.* **2011**, *2*, 340–345. [CrossRef]
23. Meng, X.; Zhao, C.; Xu, B.; Wang, P.; Liu, J. Effects of the annealing temperature on the structure and up-conversion photoluminescence of ZnO film. *J. Mater. Sci. Technol.* **2018**, *34*, 2392–2397. [CrossRef]
24. Chaitra, U.; Kekuda, D.; Rao, K.M. Effect of annealing temperature on the evolution of structural, microstructural, and optical properties of spin coated ZnO thin films. *Ceram. Int.* **2017**, *43*, 7115–7122. [CrossRef]
25. Raoufi, D.; Raoufi, T. The effect of heat treatment on the physical properties of sol–gel derived ZnO thin films. *Appl. Surf. Sci.* **2009**, *255*, 5812–5817. [CrossRef]
26. O'Brien, S.; Nolan, M.G.; Çopuroglu, M.; Hamilton, J.A.; Povey, I.; Pereira, L.; Martins, R.; Fortunato, E.; Pemble, M. Zinc oxide thin films: Characterization and potential applications. *Thin Solid Films* **2010**, *518*, 4515–4519. [CrossRef]
27. Arif, M.; Sanger, A.; Vilarinho, P.M.; Cho, M.H. Effect of annealing temperature on structural and optical properties of sol-gel-derived ZnO thin films. *J. Electron. Mater.* **2018**, *47*, 3678–3684. [CrossRef]
28. Guillemin, S.; Rapenne, L.; Sarigiannidou, E.; Donatini, F.; Consonni, V.; Bremond, G. Identification of extended defect and interface related luminescence lines in polycrystalline ZnO thin films grown by sol–gel process. *RSC Adv.* **2016**, *6*, 44987–44992. [CrossRef]
29. Sato, M.; Hara, H.; Nishide, T.; Sawada, Y. A water-resistant precursor in a wet process for TiO_2 thin film formation. *J. Mater. Chem.* **1996**, *6*, 1767. [CrossRef]
30. Likius, D.S.; Nagai, H.; Aoyama, S.; Mochizuki, C.; Hara, H.; Baba, N.; Sato, M. Percolation threshold for electrical resistivity of Ag-nanoparticle/titania composite thin films fabricated using molecular precursor method. *J. Mater. Sci.* **2012**, *47*, 3890–3899. [CrossRef]
31. Nagai, H.; Sato, M. Molecular Precursor Method for Fabricating p-Type Cu_2O and Metallic Cu Thin Films. In *Modern Technologies for Creating the Thin-film Systems and Coatings*; IntechOpen: London, UK, 2017.
32. Daniel, L.S.; Nagai, H.; Sato, M. Absorption spectra and photocurrent densities of Ag nanoparticle/TiO_2 composite thin films with various amounts of Ag. *J. Mater. Sci.* **2013**, *48*, 7162–7170. [CrossRef]
33. Mashiyama, Y.; Yoshioka, K.; Komiyama, S.; Nomura, H.; Adachi, S.; Sato, M.; Honda, T. Fabrication of MgZnO films by molecular precursor method and their application to UV-transparent electrodes. *Phys. Status Solidi c* **2009**, *6*, 596–598. [CrossRef]
34. Taka, D.; Onuma, T.; Shibukawa, T.; Nagai, H.; Yamaguchi, T.; Jang, J.-S.; Sato, M.; Honda, T. Fabrication of Ag dispersed ZnO films by molecular precursor method and application in GaInN blue LED. *Phys. Status Solidi a* **2016**, *214*, 1600598. [CrossRef]
35. Xu, J.; Pan, Q.; Shun, Y.; Tian, Z. Grain size control and gas sensing properties of ZnO gas sensor. *Sensors Actuators B Chem.* **2000**, *66*, 277–279. [CrossRef]
36. Ohyama, M.; Kozuka, H.; Yoko, T.; Sakka, S. Preparation of ZnO films with preferential orientation by sol-gel method. *J. Ceram. Soc. Jpn.* **1996**, *104*, 296–300. [CrossRef]
37. Nishio, K.; Miyake, S.; Sei, T.; Watanabe, Y.; Tsuchiya, T. Preparation of highly oriented thin film exhibiting transparent conduction by the sol-gel process. *J. Mater. Sci.* **1996**, *31*, 3651–3656. [CrossRef]
38. Chakrabarti, S.; Ganguli, D.; Chaudhuri, S. Substrate dependence of preferred orientation in sol–gel-derived zinc oxide films. *Mater. Lett.* **2004**, *58*, 3952–3957. [CrossRef]
39. Nesheva, D.; Dzhurkov, V.; Stambolova, I.; Blaskov, V.; Bineva, I.; Moreno, J.M.C.; Preda, S.; Gartner, M.; Hristova-Vasileva, T.; Shipochka, M. Surface modification and chemical sensitivity of sol gel deposited nanocrystalline ZnO films. *Mater. Chem. Phys.* **2018**, *209*, 165–171. [CrossRef]
40. Srinivasan, G.; Gopalakrishnan, N.; Yu, Y.; Kesavamoorthy, R.; Kumar, J. Influence of post-deposition annealing on the structural and optical properties of ZnO thin films prepared by sol–gel and spin-coating method. *Superlattices Microstruct.* **2008**, *43*, 112–119. [CrossRef]
41. Smirnov, M.; Baban, C.; Rusu, G. Structural and optical characteristics of spin-coated ZnO thin films. *Appl. Surf. Sci.* **2010**, *256*, 2405–2408. [CrossRef]
42. Malek, M.; Mamat, M.H.; Khusaimi, Z.; Sahdan, M.Z.; Musa, M.; Zainun, A.; Suriani, A.; Sin, N.M.; Hamid, S.B.A.; Rusop, M. Sonicated sol–gel preparation of nanoparticulate ZnO thin films with various deposition speeds: The highly preferred c-axis (002) orientation enhances the final properties. *J. Alloy. Compd.* **2014**, *582*, 12–21. [CrossRef]

43. Nagai, H.; Sato, M. Heat Treatment in Molecular Precursor Method for Fabricating Metal Oxide Thin Films. In *Heat Treatment—Conventional and Novel Applications*; IntechOpen: London, UK, 2012; p. 13.
44. Nagai, H.; Sato, M. The Science of Molecular Precursor Method. In *Advanced Coating Materials*; Wiley: Hoboken, NJ, USA, 2018; pp. 1–27.
45. Salam, S.; Islam, M.; Akram, A. Sol–gel synthesis of intrinsic and aluminum-doped zinc oxide thin films astransparent conducting oxides for thin film solar cells. *Thin Solid Films* **2013**, *529*, 242–247. [CrossRef]
46. Tiwale, N.; Senanayak, S.P.; Rubio-Lara, J.; Alaverdyan, Y.; Welland, M.E. Optimization of transistor characteristics and charge transport in solution processed ZnO thin films grown from zinc neodecanoate. *Electron. Mater. Lett.* **2019**, *15*, 702–711. [CrossRef]
47. Chen, M.-Z.; Chen, W.-S.; Jeng, S.-C.; Yang, S.-H.; Chung, Y.-F. Liquid crystal alignment on zinc oxide nanowire arrays for LCDs applications. *Opt. Express* **2013**, *21*, 29277–29282. [CrossRef] [PubMed]
48. Kabir, M.H.; Ali, M.M.; Kaiyum, M.A.; Rahman, M.S. Effect of annealing temperature on structural morphological and optical properties of spray pyrolized Al-doped ZnO thin films. *J. Phys. Commun.* **2019**, *3*, 105007. [CrossRef]
49. Likhachev, D.V.; Malkova, N.; Poslavsky, L. Modified Tauc–Lorentz dispersion model leading to a more accurate representation of absorption features below the bandgap. *Thin Solid Films* **2015**, *589*, 844–851. [CrossRef]
50. Navin, K.; Kurchania, R. Structural, morphological and optical studies of ripple-structured ZnO thin films. *Appl. Phys. A* **2015**, *121*, 1155–1161. [CrossRef]
51. Soleimanian, V.; Aghdaee, S.R. The influence of annealing temperature on the slip plane activity and optical properties of nanostructured ZnO films. *Appl. Surf. Sci.* **2011**, *258*, 1495–1504. [CrossRef]
52. Ong, H.C.; Zhu, A.X.E.; Du, G.T. Dependence of the excitonic transition energies and mosaicity on residual strain in ZnO thin films. *Appl. Phys. Lett.* **2002**, *80*, 941–943. [CrossRef]
53. Akhtar, M.S.; Riaz, S.; Noor, R.; Naseem, S. Optical and Structural Properties of ZnO Thin Films for Solar Cell Applications. *Adv. Sci. Lett.* **2013**, *19*, 834–838. [CrossRef]
54. Wahab, H.; Salama, A.; El Saeid, A.A.; Nur, O.; Willander, M.; Battisha, I. Optical, structural and morphological studies of (ZnO) nano-rod thin films for biosensor applications using sol gel technique. *Results Phys.* **2013**, *3*, 46–51. [CrossRef]
55. Samanta, P.K. Weak quantum confinement in ZnO nanorods: A one dimensional potential well approach. *Opt. Photon. Lett.* **2011**, *4*, 35–45. [CrossRef]
56. Gonzalez, A.E.J.; Urueta, J.A.Z.; Suarez-Parra, R. Optical and structural characteristics of aluminum-doped ZnO thin films prepared by sol-gel technique. *J. Cryst. Growth* **1998**, *192*, 430–438. [CrossRef]
57. Kumari, L.; Li, W.; Vannoy, C.; Leblanc, R.M.; Wang, D. Zinc oxide micro- and nanoparticles: Synthesis, structure and optical properties. *Mater. Res. Bull.* **2010**, *45*, 190–196. [CrossRef]
58. Ramanathan, S.; Patibandla, S.; Bandyopadhyay, S.; Edwards, J.D.; Anderson, J. Fluorescence and infrared spectroscopy of electrochemically self assembled ZnO nanowires: Evidence of the quantum confined Stark effect. *J. Mater. Sci. Mater. Electron.* **2006**, *17*, 651–655. [CrossRef]

© 2020 by the authors. Licensee MDPI, Basel, Switzerland. This article is an open access article distributed under the terms and conditions of the Creative Commons Attribution (CC BY) license (http://creativecommons.org/licenses/by/4.0/).

Article

Growth and Photocatalytic Properties of Gallium Oxide Films Using Chemical Bath Deposition

Che-Yuan Yeh [1,†], Yi-Man Zhao [1,2,†], Hui Li [1,2], Fei-Peng Yu [3], Sam Zhang [2,*] and Dong-Sing Wuu [1,4,*]

1. Department of Materials Science and Engineering, National Chung Hsing University, No. 145, Xingda Rd., Taichung 40227, Taiwan; w105066022@mail.nchu.edu.tw (C.-Y.Y.); zym19920609@163.com (Y.-M.Z.); lihui@kaist.ac.kr (H.L.)
2. Center for Advanced Thin Films and Devices, School of Materials and Energy, Southwest University, No. 2, Tiansheng Rd., Chongqing 400715, China
3. Department of Materials Science and Engineering, Da-Yeh University, No. 168, University Rd., Changhua 51591, Taiwan; pon.pon@ms25.url.com.tw
4. Innovation and Development Center of Sustainable Agriculture, National Chung Hsing University, No. 145, Xingda Rd., Taichung 40227, Taiwan
* Correspondence: SamZhang@swu.edu.cn (S.Z.); dsw@nchu.edu.tw (D.-S.W.)
† These authors contributed equally to this work.

Received: 4 October 2019; Accepted: 25 October 2019; Published: 27 October 2019

Abstract: Gallium oxide (Ga_2O_3) thin films were fabricated on glass substrates using a combination of chemical bath deposition and post-annealing process. From the field-emission scanning electron microscopy and x-ray diffraction results, the GaOOH nanorods precursors with better crystallinity can be achieved under higher concentrations (≥0.05 M) of gallium nitrate ($Ga(NO_3)_3$). It was found that the GaOOH synthesized from lower $Ga(NO_3)_3$ concentration did not transform into α-Ga_2O_3 among the annealing temperatures used (400–600 °C). Under higher $Ga(NO_3)_3$ concentrations (≥0.05 M) with higher annealing temperatures (≥500 °C), the GaOOH can be transformed into the Ga_2O_3 film successfully. An α-Ga_2O_3 sample synthesized in a mixed solution of 0.075 M $Ga(NO_3)_3$ and 0.5 M hexamethylenetetramine exhibited optimum crystallinity after annealing at 500 °C, where the α-Ga_2O_3 nanostructure film showed the highest aspect ratio of 5.23. As a result, the photodegeneration efficiencies of the α-Ga_2O_3 film for the methylene blue aqueous solution can reach 90%.

Keywords: Ga_2O_3; GaOOH; chemical bath deposition; photocatalytic property

---

1. Introduction

Environmental protection, especially against water pollution, has generated public interest in recent years, and the application of nanotechnology in this field has become a hot topic. Photocatalysis using natural sunlight as a clean energy source, has been extensively studied for environmental remediation since its discovery [1–4]. Methylene blue (MB), a common dye, has been widely used for the commercial products such as the dyeing paper, linen, silk fabric, bamboo, wood, etc. The discharge of untreated MB into the wastewater can lead to serious pollution problems, making degradation and decolorization of MB is one of the important targets of dyeing wastewater treatment [5,6].

Metal oxides, such as titanium dioxide, zinc oxide, and gallium oxide (Ga_2O_3), are considered promising for photocatalytic applications due to their excellent physical and chemical properties [7–11]. Among these metal oxides, Ga_2O_3, with various polymorphs (α-, β-, γ-, δ-, and ε-) can be synthesized at different temperatures [12,13]. Gallium oxides, which have wide bandgap ranging from 4.2 to 4.9 eV, are particularly important semiconductor materials [11,14]. The Ga_2O_3 material can be used for various device applications, such as gas sensors [15], catalysis [16], power and high voltage electronic devices [13] because it has unique conductive properties and is electrically and thermodynamically stable [17–19]. Various polymorphs of Ga_2O_3 as the photocatalysts in the degradation of organic dyes have been studied previously [11,20–22]. Reddy et al. [11] have reported that the photodegeneration efficiencies of α-Ga_2O_3 and β-Ga_2O_3 nanorods for the rhodamine B aqueous solution were 62% and 79%, respectively. The photocatalytic activity of α-Ga_2O_3 in the malachite green degradation was studied by Rodríguez et al. [20]. Zhao et al. [21], have demonstrated that the photodegeneration efficiency of β-Ga_2O_3 nanorods in the perfluorooctanic acid can reach 98.8%. In addition, the photocatalytic activity of the γ-Ga_2O_3 in rhodamine 590 degradation has also been investigated by Jin et al. [22].

Recently, various methods have been used to synthesize Ga_2O_3, including hydrothermal techniques [11,23], microwave-assisted hydrothermal methods [21], sol-gel methods [24], spray pyrolysis processes [25], and chemical vapor deposition [26]. The advantages of CBD for the synthesis of metal oxide thin films are based on the relatively low cost and convenience for deposition over a large area [27]. However, there are very few reports on the synthesis of Ga_2O_3 thin films via CBD. In this study, the Ga_2O_3 thin films were synthesized on the glass substrates using CBD with various post annealing treatments. The chemistry of the synthesis process can be summarized as follows [20,28–30]:

$$(CH_2)_6N_4 + 6H_2O \rightarrow 4NH_3 + 6HCHO \quad (1)$$

$$NH_3 + H_2O \rightarrow NH^{4+} + OH \quad (2)$$

$$3OH^- + Ga^{3+} \rightarrow Ga(OH)_3 \quad (3)$$

$$Ga(OH)_3 \rightarrow GaOOH + H_2O \quad (4)$$

2. Experimental

Ga_2O_3 thin films were prepared on glass substrates using CBD. Analytical grade gallium (III) nitrate hydrate ($Ga(NO_3)_3 \cdot nH_2O$, 99.9%) and hexamethylenetetramine (HMT, 99.9%) were used as starting materials. The glass substrates were cleaned with acetone, methanol, and 10% hydrofluoric acid for 10 min respectively, then rinsed with deionized (DI) water. Three growth solutions were prepared by dissolving 0.025 mol, 0.05 moL, or 0.075 mol $Ga(NO_3)_3$ and 0.5 mol HMT in 1 L of DI water. To obtain the GaOOH precursor, each aqueous solution was placed in a beaker with a glass substrate and vigorously stirred at 500 rpm with a magnetic stirrer for 5 h at 95 °C. The precursor was then dried at 70 °C for 5 min and allowed to cool naturally to room temperature. The as-prepared GaOOH thin films were annealed in a furnace tube at either 400 °C, 500 °C, or 600 °C for 3 h to form Ga_2O_3 thin films.

The crystal structures of the as-prepared samples were characterized using a high resolution X-ray diffraction system (HR-XRD, X'Pert Pro MRD, PANanalytical, Almelo, Nederland). A JSM-6700F field-emission scanning electron microscope (FE-SEM, JEOL, Tokyo, Japan) equipped with an energy dispersive spectrometer (EDS) was used to analyze the morphologies of the samples and their elemental distributions. FT-IR spectra were collected in the range from 4000 to 500 cm^{-1} with a Vertex 80v Fourier transform infrared spectrometer (Bruker, Billerica, Massachusetts, MA, USA). The photocatalytic properties of the Ga_2O_3 thin films in aqueous MB solutions were evaluated during and after irradiation under an 8 W UV lamp at wavelengths from 100 to 280 nm. Thin films with dimensions of 1 cm × 1 cm were first placed in aqueous solutions containing 5 ppm MB. The solutions were then irradiated for 5 h at room temperature in an otherwise dark environment. Photodegradation efficiency was determined through UV-visible spectrophotometric analysis (U3010, Hitachi, Tokyo, Japan)

3. Results and Discussion

The FE-SEM images of the GaOOH nanostructures obtained via CBD at various $Ga(NO_3)_3$ concentrations are shown in Figure 1, where the length, width, and the aspect ratio of at least eight GaOOH nanorods were measured. It was found that the $Ga(NO_3)_3$ concentration significantly influenced the synthesis of GaOOH nanorods and the formation of Ga_2O_3 thin films during the annealing process. The GaOOH crystalline quality increased as the $Ga(NO_3)_3$ concentration increased from 0.025 M (Figure 1a,d) to 0.075 M (Figure 1c,f). The average length, width and the aspect ratio of the GaOOH nanorods prepared from 0.025 M $Ga(NO_3)_3$ were 0.88 µm, 0.27 µm and 3.23, respectively (Figure 1d). The average length, width and the aspect ratio of the GaOOH nanorods prepared from 0.05 M $Ga(NO_3)_3$ were 0.92 µm, 0.25 µm and 3.76, respectively (Figure 1e). When the $Ga(NO_3)_3$ concentration increased to 0.075 M, the average length, width and the aspect ratio of the GaOOH nanorods increased to 1.24 µm, 0.25 µm and 5.06, respectively (Figure 1f). These indicated that the higher concentration of $Ga(NO_3)_3$ yielded the GaOOH nanorods with better crystallinity.

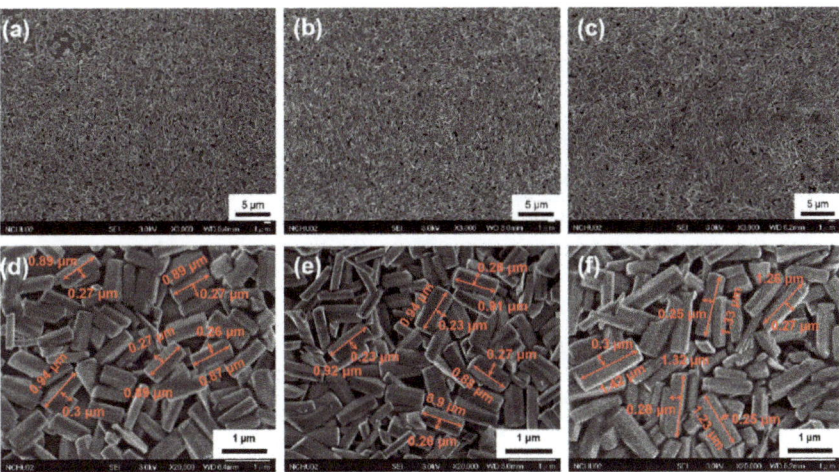

Figure 1. FE-SEM images of GaOOH prepared from $Ga(NO_3)_3$ at concentrations ranging from 0.025 M to 0.075 M: (**a**) 0.025 M; (**b**) 0.05 M; (**c**) 0.075 M. Their corresponding enlarged micrographs are shown in (**d**–**f**), respectively.

Figure 2a shows the XRD pattern of the as-prepared GaOOH obtained at each $Ga(NO_3)_3$ concentration. The full width at half maximum (FWHM) of each (110) crystal plane diffraction peak was calculated from XRD data as shown in Figure 2b. The XRD patterns indicated the as-prepared GaOOH had an orthorhombic structure (JCPDS no. 01–180). The sharp diffraction peak of the (110) crystal plane appeared at (2θ) 21.4° in the pattern of each sample. This indicated that GaOOH crystal growth proceeded preferentially in the [001] direction at each $Ga(NO_3)_3$ concentration. The diffraction peaks of the (130), (111), and (240) crystal planes of GaOOH were observed at (2θ) 33.7°, 37.2°, and 54.02°, respectively. The peaks at 21.4° in the patterns of GaOOH nanorods prepared using 0.025 M, 0.05 M, and 0.075 M $Ga(NO_3)_3$ had FWHMs of 878, 759.7, and 745.1 arcsec, respectively (Figure 2b). The FWHM of this peak was related to the crystallinity of the GaOOH nanorods and indicated that the crystallinity of the nanorods prepared with 0.075 M $Ga(NO_3)_3$ was higher than that of the other GaOOH samples. The XRD results agreed well with the crystallinity trend of the GaOOH nanorods deposited from various $Ga(NO_3)_3$ concentrations as presented in Figure 1.

An FTIR spectra of the as-prepared GaOOH nanorods synthesized at each $Ga(NO_3)_3$ concentration is shown in Figure 3. It is known that the peak at ~3222 cm^{-1} in the FTIR spectra of GaOOH nanorods was assigned to the H–OH stretching vibration, and the stretching vibration of O–H bonds was observed around 1323 cm^{-1} due to the adsorption of water molecules [11,31]. The peak at ~750 cm^{-1} was assigned to the Ga-O-H stretching vibration [20]. It is worthy to mention that the Ga-O-H bond will be enhanced and shifted from ~750 to ~820 cm^{-1} as the GaOOH content increased, [31,32]. This confirmed that the GaOOH nanostructure thin films were successfully grown on the glass substrates via CBD.

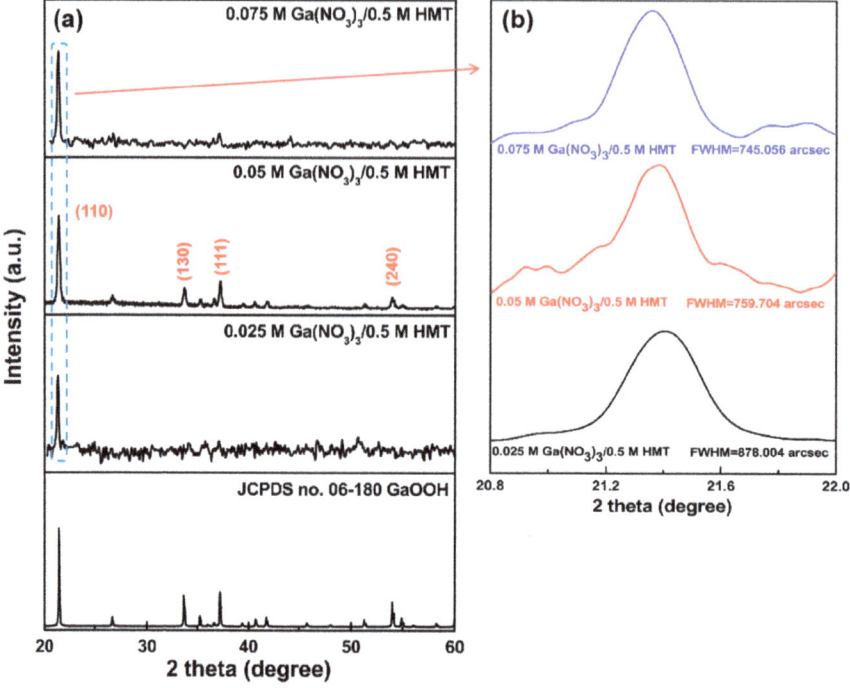

Figure 2. (a) XRD patterns of as-prepared GaOOH nanorods obtained from Ga(NO$_3$)$_3$ at concentrations of 0.025 M, 0.05 M, and 0.075 M. (b) FWHMs of the GaOOH (110) crystal plane diffraction peaks shown in (a).

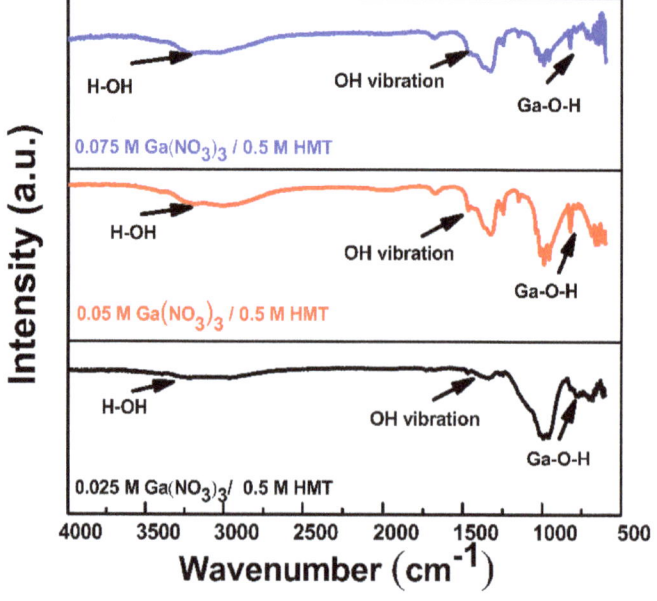

Figure 3. FTIR spectra of GaOOH prepared from Ga(NO$_3$)$_3$ at various concentrations.

Figure 4 shows the XRD patterns of Ga_2O_3 in the films after annealing for 3 h at various temperatures from 400 to 600 °C. As shown in Figure 4a, the α-Ga_2O_3 polymorph (JCPDS no. 06–0503) was not indicated in any of the samples annealed at 400 °C. Possibly, the GaOOH could not change to α-Ga_2O_3 at such as a low temperature [12]. It was also found that the GaOOH synthesized from 0.025 M Ga(NO_3)$_3$ did not transform into α-Ga_2O_3 among these annealing temperatures. However, the transformation of as-prepared GaOOH into orthorhombic α-Ga_2O_3 occurred during annealing at 500 and 600 °C. The sharp diffraction peaks of the (104) and (110) α-Ga_2O_3 crystal planes were observed at (2θ) 33.78° and 36.03°, respectively. Table 1 shows the FWHM data of the (104) α-Ga_2O_3 diffraction peaks obtained from 500 and 600 °C-annealed GaOOH samples with 0.05 and 0.075 M Ga(NO_3)$_3$ concentrations. Obviously, the CBD-GaOOH sample with 0.075 M Ga(NO_3)$_3$ concentration and post thermal treatment at 500 °C can result in higher crystallinity of α-Ga_2O_3

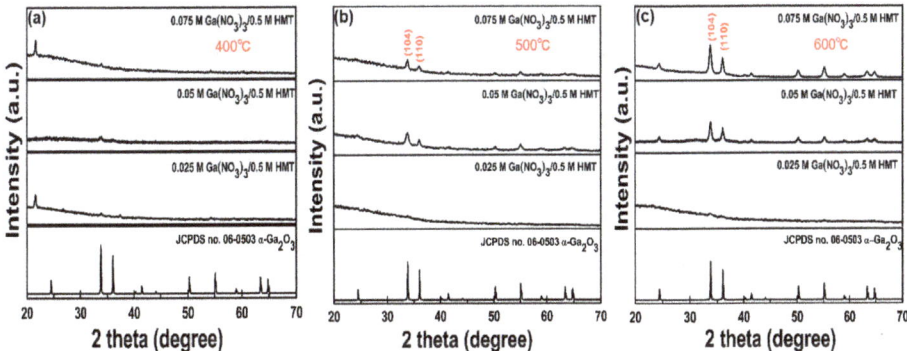

Figure 4. XRD patterns of Ga_2O_3 obtained from thermal-treated GaOOH samples for 3 h at various temperatures (**a**) 400 °C, (**b**) 500 °C and (**c**) 600 °C.

Table 1. The FWHM data of (104) α-Ga_2O_3 diffraction peaks obtained from post-annealed GaOOH samples with 0.05 and 0.075 M Ga(NO_3)$_3$ concentrations.

Annealed Temperature	0.05 M Ga(NO_3)$_3$	0.075 M Ga(NO_3)$_3$
500 °C	2153.5 arcsec	1567.4 arcsec
600 °C	1715 arcsec	1700.6 arcsec

Figure 5 shows the corresponding FE-SEM micrograph of the α-Ga_2O_3 sample as tabulated in Table 1. After annealing at 500 °C, the average length, width and the aspect ratio of the Ga_2O_3 nanorods obtained from CBD-GaOOH with 0.05 M Ga(NO_3)$_3$ were 0.94 μm, 0.26 μm and 3.63, respectively (Figure 5a). When the annealing temperature increased to 600 °C, the average length, width and the aspect ratio of the Ga_2O_3 nanorods were 0.89 μm, 0.26 μm, and 3.45, respectively (Figure 5b). Besides, the average length, width and the aspect ratio of the Ga_2O_3 nanorods obtained from 500 °C-annealed GaOOH with 0.075 M Ga(NO_3)$_3$ were 1.16 μm, 0.23 μm and 5.23, respectively (Figure 5c). When the annealing temperature increased to 600 °C, the average length, width and the aspect ratio of the Ga_2O_3 nanorods were 1.22 μm, 0.26 μm, and 4.68, respectively (Figure 5d). Here, all the data of length, width, and the aspect ratio shown in the micrographs were measured at least eight times. It becomes clear that the higher annealing temperature (i.e., 600 °C) will lead to the slight decrease in the aspect ratio of Ga_2O_3 nanorods. As a result, these nanorods were tightly packed together.

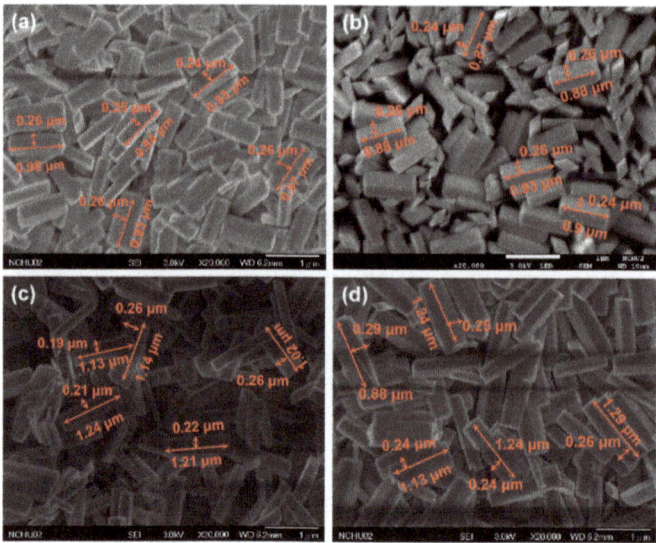

Figure 5. FE-SEM Ga$_2$O$_3$ obtained after annealing GaOOH at 500 °C: (**a**) 0.05 M; (**c**) 0.075 M; and 600 °C: (**b**) 0.05 M; (**d**) 0.075 M.

The elemental compositions of the annealed samples were analyzed via EDS. The EDS spectra of samples prepared using Ga(NO$_3$)$_3$ at various concentrations after annealing at 500 °C are shown in Figure 6. The Si peaks in the EDS spectra were attributed to silicon in the glass substrates. The at.% of Ga in the sample prepared from 0.025 M Ga(NO$_3$)$_3$ was only 1.92% (Figure 6a). When the Ga(NO$_3$)$_3$ concentration increased to 0.05 M, the content of Ga atoms in the sample increased to 16.56% (Figure 6b). Moreover, the at.% of Ga in the sample prepared from 0.075 M Ga(NO$_3$)$_3$ was 30.35% (Figure 6c). From the XRD results as described in Figure 4, the CBD-GaOOH prepared under the lower concentration (0.025 M) cannot be transformed into the α-Ga$_2$O$_3$ regardless of the annealing temperature. It may be due to the fact that the 0.025 M Ga(NO$_3$)$_3$ cannot provide enough Ga content to transform into α-Ga$_2$O$_3$.

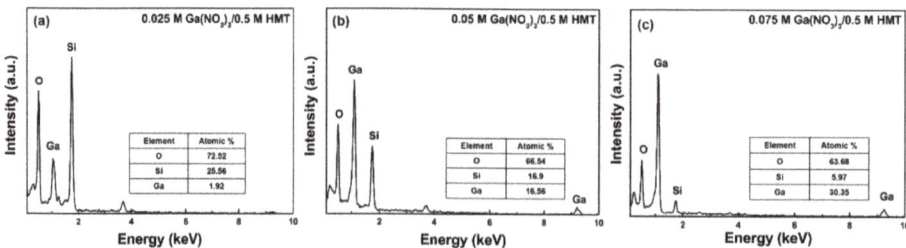

Figure 6. EDS spectra of annealed samples synthesized using (**a**) 0.025 M Ga(NO$_3$)$_3$, (**b**) 0.05 M Ga(NO$_3$)$_3$ and (**c**) 0.075 M Ga(NO$_3$)$_3$. The samples were annealed for 3 h at 500 °C.

The photocatalytic properties of the α-Ga$_2$O$_3$ thin films in the MB solutions were examined under irradiation with UV light. The Ga$_2$O$_3$ shows the photocatalytic activity under deep-UV absorption owing to its high bandgap. In general, when the photon was absorbed into photocatalyst, the electron-hole pairs generated on the surface of Ga$_2$O$_3$ were used to the degradation of the MB solution. After interacting with aqueous media, these electrons and holes generate hydroxyl ions, which play the significant role in the degradation process. These hydroxyl ions have strong oxidation capabilities as they can mineralize most of organic compounds. Moreover, the photocatalytic activities intimately depend on the crystallinity and surface area. From the XRD (Figure 4) and FWHMs analyses (Table 1), the good crystallinity of α-Ga$_2$O$_3$ sample prepared using 0.075 M Ga(NO$_3$)$_3$ can bring the stable photocatalytic properties as compared with those of the other α-Ga$_2$O$_3$ samples. The calculated constant reaction rates of MB photodegradation by Ga$_2$O$_3$ in the solutions (C/C$_0$) are plotted as a function of UV irradiation time in Figure 7. C$_0$ is the initial MB concentration in solution, and C is the MB concentration in the UV-irradiated solution. The rates of MB self-degradation in solutions without the catalyst are shown for comparison. Absorbance by MB was monitored at 664 nm [33]. After 5 h of UV irradiation, 82% of the MB was photodegraded by α-Ga$_2$O$_3$ obtained from GaOOH prepared from 0.05 M Ga(NO$_3$)$_3$ after annealing at 500 °C (Figure 7b). Photodegradation by α-Ga$_2$O$_3$ obtained from the GaOOH sample prepared using 0.075 M Ga(NO$_3$)$_3$ after annealing at 500 °C was particularly efficient. After UV irradiation for 5 h, 90% of the MB had been photodegraded by the catalyst (Figure 7c). As shown in Figure 7a, only 29% of the MB was photodegraded by the sample obtained from GaOOH prepared using 0.025 M Ga(NO$_3$)$_3$. Based on the XRD results (Figure 4), its low photodegradation efficiency may have been due to the lack of GaOOH transformation into α-Ga$_2$O$_3$. The photodegradation efficiencies of α-Ga$_2$O$_3$ samples obtained after annealing at 500 °C were higher than those of α-Ga$_2$O$_3$ obtained after annealing at 600 °C. This may have been due to the higher the aspect ratio of the α-Ga$_2$O$_3$ annealed at 500 °C, which can lead to more surface area under the process of the degradation of the MB solution. The photocatalytic properties of α-Ga$_2$O$_3$ observed during the degradation of MB suggested it held promise for environmental remediation applications.

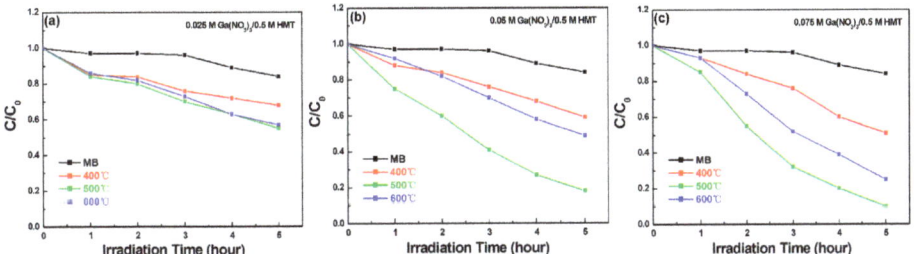

Figure 7. Calculated constant reaction rate of MB photodegradation (C/C$_0$) by Ga$_2$O$_3$ plotted as a function of UV irradiation time. The Ga$_2$O$_3$ samples were obtained following annealing of GaOOH prepared at Ga(NO$_3$)$_3$ concentrations of (**a**) 0.025 M, (**b**) 0.05 M and (**c**) 0.075 M.

4. Conclusions

GaOOH thin films were successfully grown on glass substrates via CBD at 95 °C. The as-prepared GaOOH thin films were annealed for 3 h at either 400 °C, 500 °C, or 600 °C to convert GaOOH into thin films of α-Ga$_2$O$_3$, and the crystal structures and elemental compositions were confirmed through the XRD and EDS analysis, respectively. Nanocrystalline α-Ga$_2$O$_3$ films were obtained from GaOOH prepared at Ga(NO$_3$)$_3$ concentrations at or above 0.05 M after annealing at 500 °C or 600 °C. The α-Ga$_2$O$_3$ sample obtained after annealing GaOOH prepared at a Ga(NO$_3$)$_3$ concentration of 0.075 M photodegraded 90% of the MB in a solution irradiated under a UV lamp for 5 h. These results suggest that α-Ga$_2$O$_3$ thin films can be very useful alternatives for the photocatalytic degradation of dyes during wastewater treatment.

Author Contributions: Conceptualization and supervision, D.-S.W.; methodology, C.-Y.Y.; H.L. and F.-P.Y.; validation, C.-Y.Y.; data curation, C.-Y.Y. and Y.-M.Z.; writing-review and editing; Y.-M.Z., S.Z. and D.-S.W.

Funding: This work was financially supported by the Ministry of Science and Technology of Taiwan under grant No. 108–2221-E-005–028-MY3 and in part by the "Innovation and Development Center of Sustainable Agriculture" from The Featured Areas Research Center Program within the framework of the Higher Education Sprout Project by the Ministry of Education (MOE) in Taiwan. The work at Southwest University was partly supported by Fundamental Research Funds for the Central Universities: SWU118105.

Acknowledgments: The authors would like to thank the technical support from Yi-Guo Shang in the Department of Materials Science and Engineering, National Chung Hsing University, Taiwan, especially for the setup of chemical bath deposition and photocatalytic measurement systems.

Conflicts of Interest: The authors declare that they have no conflict of interest.

References

1. Fujishima, A.; Honda, K. Electrochemical Photolysis of Water at a Semiconductor Electrode. *Nature* **1972**, *238*, 37–38. [CrossRef] [PubMed]
2. Fujishima, A. Hydrogen Production under Sunlight with an Electrochemical Photocell. *J. Electrochem. Soc.* **1975**, *122*, 1487. [CrossRef]
3. Djurišić, A.B.; Leung, Y.H.; Ng, A.M.C. Strategies for improving the efficiency of semiconductor metal oxide photocatalysis. *Mater. Horizons* **2014**, *1*, 400. [CrossRef]
4. Benjwal, P.; Kar, K.K. Simultaneous photocatalysis and adsorption based removal of inorganic and organic impurities from water by titania/activated carbon/carbonized eponocoxy namposite. *J. Environ. Chem. Eng.* **2015**, *3*, 2076–2083. [CrossRef]
5. Houas, A. Photocatalytic degradation pathway of methylene blue in water. *Appl. Catal. B Environ.* **2001**, *31*, 145–157. [CrossRef]
6. Lachheb, H.; Puzenat, E.; Houas, A.; Ksibi, M.; Elaloui, E.; Guillard, C.; Herrmann, J.-M. Photocatalytic degradation of various types of dyes (Alizarin S, Crocein Orange G, Methyl Red, Congo Red, Methylene Blue) in water by UV-irradiated titania. *Appl. Catal. B Environ.* **2002**, *39*, 75–90. [CrossRef]
7. Tan, L.-L.; Ong, W.-J.; Chai, S.-P.; Goh, B.T.; Mohamed, A.R. Visible-light-active oxygen-rich TiO_2 decorated 2D graphene oxide with enhanced photocatalytic activity toward carbon dioxide reduction. *Appl. Catal. B Environ.* **2015**, *179*, 160–170. [CrossRef]
8. Liu, L.; Gao, F.; Zhao, H.; Li, Y. Tailoring Cu valence and oxygen vacancy in Cu/TiO_2 catalysts for enhanced CO_2 photoreduction efficiency. *Appl. Catal. B Environ.* **2013**, *134*, 349–358. [CrossRef]
9. Núñez, J.; O'Shea, V.A.D.L.P.; Jana, P.; Coronado, J.M.; Serrano, D.P.; O'Shea, V.A.D.L.P. Effect of copper on the performance of ZnO and ZnO1−xNx oxides as CO_2 photoreduction catalysts. *Catal. Today* **2013**, *209*, 21–27. [CrossRef]
10. Pan, Y.-X.; Sun, Z.-Q.; Cong, H.-P.; Men, Y.-L.; Xin, S.; Song, J.; Yu, S.-H. Photocatalytic CO_2 reduction highly enhanced by oxygen vacancies on Pt-nanoparticle-dispersed gallium oxide. *Nano Res.* **2016**, *9*, 1689–1700. [CrossRef]
11. Reddy, L.S.; Ko, Y.H.; Yu, J.S. Hydrothermal Synthesis and Photocatalytic Property of β-Ga_2O_3 Nanorods. *Nanoscale Res. Lett.* **2015**, *10*, 364. [CrossRef] [PubMed]
12. Roy, R.; Hill, V.G.; Osborn, E.F. Polymorphism of Ga_2O_3 and the System Ga_2O_3—H_2O. *J. Am. Chem. Soc.* **1952**, *74*, 719–722. [CrossRef]
13. Stepanov, S.I.; Nikolaev, V.I.; Bougrov, V.E.; Romanov, A.E. Gallium oxide: Properties and applications-a review. *Mater. Sci.* **2016**, *44*, 63–86.
14. Liu, X.; Qiu, G.; Zhao, Y.; Zhang, N.; Yi, R. Gallium oxide nanorods by the conversion of gallium oxide hydroxide nanorods. *J. Alloys Compd.* **2007**, *439*, 275–278. [CrossRef]
15. Pilliadugula, R.; Krishnan, N. Gas sensing performance of GaOOH and β-Ga_2O_3 synthesized by hydrothermal method: A comparison. *Mater. Res. Express* **2019**, *6*, 025027. [CrossRef]
16. Nakagawa, K.; Kajita, C.; Okumura, K.; Ikenaga, N.-O.; Nishitani-Gamo, M.; Ando, T.; Kobayashi, T.; Suzuki, T. Role of Carbon Dioxide in the Dehydrogenation of Ethane over Gallium-Loaded Catalysts. *J. Catal.* **2001**, *203*, 87–93. [CrossRef]
17. Kuperberg, J.M.; Réti, F.; Miró, J.; Herndon, R.C.; Hajnal, Z.; Kiss, G.; Deák, P. Role of oxygen vacancy defect states in the n -type conduction of β-Ga_2O_3. *J. Appl. Phys.* **1999**, *86*, 3792–3796.
18. Wu, X.; Song, W.; Huang, W.; Pu, M.; Zhao, B.; Sun, Y.; Du, J. Crystalline gallium oxide nanowires: Intensive blue light emitters. *Chem. Phys. Lett.* **2000**, *328*, 5–9. [CrossRef]
19. Binet, L.; Gourier, D. ORIGIN OF THE BLUE LUMINESCENCE OF β-Ga_2O_3. *J. Phys. Chem. Solids* **1998**, *59*, 1241–1249. [CrossRef]

20. Rodríguez, C.I.M.; Álvarez, M.; Ángel, L.; Rivera, J.D.J.F.; Arízaga, G.G.C.; Michel, C.R. α-Ga$_2$O$_3$ as a Photocatalyst in the Degradation of Malachite Green. *ECS J. Solid State Sci. Technol.* **2019**, *8*, Q3180–Q3186. [CrossRef]
21. Zhao, B.X.; Li, X.; Yang, L.; Wang, F.; Li, J.C.; Xia, W.X.; Li, W.J.; Zhou, L.; Zhao, C.L. β-Ga$_2$O$_3$ nanorod synthesis with a one-step microwave irradiation hydrothermal method and its efficient photocatalytic degradation for perfluorooctanoic acid. *Photochem. Photobiol.* **2015**, *91*, 42–47. [CrossRef] [PubMed]
22. Jin, S.; Lu, W.; Stanish, P.C.; Radovanovic, P.V. Compositional control of the photocatalytic activity of Ga$_2$O$_3$ nanocrystals enabled by defect-induced carrier trapping. *Chem. Phys. Lett.* **2018**, *706*, 509–514. [CrossRef]
23. Wang, Y.; Li, N.; Duan, P.; Sun, X.; Chu, B.; He, Q. Properties and Photocatalytic Activity of β-Ga$_2$O$_3$ Nanorods under Simulated Solar Irradiation. *J. Nanomater.* **2015**, *2015*, 126.
24. Minami, T. Oxide thin-film electroluminescent devices and materials. *Solid-State Electron.* **2003**, *47*, 2237–2243. [CrossRef]
25. Ji, Z.; Du, J.; Fan, J.; Wang, W. Gallium oxide films for filter and solar-blind UV detector. *Opt. Mater.* **2006**, *28*, 415–417. [CrossRef]
26. Oshima, Y.; Víllora, E.G.; Shimamura, K. Quasi-heteroepitaxial growth of β-Ga$_2$O$_3$ on off-angled sapphire (0001) substrates by halide vapor phase epitaxy. *J. Cryst. Growth* **2015**, *410*, 53–58. [CrossRef]
27. Pawar, S.; Pawar, B.; Kim, J.; Joo, O.-S.; Lokhande, C. Recent status of chemical bath deposited metal chalcogenide and metal oxide thin films. *Curr. Appl. Phys.* **2011**, *11*, 117–161. [CrossRef]
28. Zainelabdin, A.; Zaman, S.; Amin, G.; Nur, O.; Willander, M. Deposition of well-aligned ZnO nanorods at 50 °C on metal, semiconducting polymer, and copper oxides substrates and their structural and optical properties. *Cryst. Growth Des.* **2010**, *10*, 3250–3256. [CrossRef]
29. Errico, V.; Arrabito, G.; Plant, S.R.; Medaglia, P.G.; Palmer, R.E.; Falconi, C. Chromium inhibition and sizeselected Au nanocluster catalysis for the solution growth of lowdensity ZnO nanowires. *Sci. Rep.* **2015**, *5*, 12336. [CrossRef]
30. Arrabitoa, G.; Erricoa, V.; Zhang, Z.M.; Han, W.H.; Falconia, C. Nanotransducers on printed circuit boards by rational design of highdensity, long, thin and untapered ZnO nanowires. *Nano Energy* **2018**, *46*, 54–62. [CrossRef]
31. Zhao, Y.; Frost, R.L.; Yang, J.; Martens, W.N. Size and Morphology Control of Gallium Oxide Hydroxide GaO(OH), Nano- to Micro-Sized Particles by Soft-Chemistry Route without Surfactant. *J. Phys. Chem. C* **2008**, *112*, 3568–3579. [CrossRef]
32. Huang, C.-C.; Yeh, C.-S.; Ho, C.-J. Laser Ablation Synthesis of Spindle-like Gallium Oxide Hydroxide Nanoparticles with the Presence of Cationic Cetyltrimethylammonium Bromide. *J. Phys. Chem. B* **2004**, *108*, 4940–4945. [CrossRef]
33. Yao, J.; Wang, C. Decolorization of Methylene Blue with TiO$_2$ Sol via UV Irradiation Photocatalytic Degradation. *Int. J. Photoenergy* **2010**, *2010*, 643182. [CrossRef]

© 2019 by the authors. Licensee MDPI, Basel, Switzerland. This article is an open access article distributed under the terms and conditions of the Creative Commons Attribution (CC BY) license (http://creativecommons.org/licenses/by/4.0/).

Article

Effect of Nitrogen Flow in Hydrogen/Nitrogen Plasma Annealing on Aluminum-Doped Zinc Oxide/Tin-Doped Indium Oxide Bilayer Films Applied in Low Emissivity Glass

Shang-Chou Chang [1,2,*] and Huang-Tian Chan [2]

[1] Department of Electrical Engineering, Kun Shan University, No.195, Kunda Road, Yong-kang District, Tainan City 71070, Taiwan
[2] Green Energy Technology Research Center, Kun Shan University, No.195, Kunda Road, Yong-kang District, Tainan City 71070, Taiwan; g5eek79@gmail.com
* Correspondence: jchang@mail.ksu.edu.tw; Tel.: +886-6-2051512

Received: 28 May 2019; Accepted: 13 June 2019; Published: 17 June 2019

Abstract: Low emissivity glass (low-e glass), which is often used in energy-saving buildings, has high thermal resistance and visible light transmission. Heavily doped wide band gap semiconductors like aluminum-doped zinc oxide (AZO) and tin-doped indium oxide (ITO) have these properties, especially after certain treatment. In our experiments, in-line sputtered AZO and ITO bilayer (AZO/ITO) films on glass substrates were prepared first. The deposition of AZO/ITO films was following by annealing in hydrogen/nitrogen (H_2/N_2) plasma with different N_2 flows. The structure and optical and electrical properties of AZO/ITO films were surveyed. Experiment results indicated that N_2 flow in H_2/N_2 plasma annealing of AZO/ITO films slightly modified the structure and electrical properties of AZO/ITO films. The X-ray diffraction peak corresponding to zinc oxide (002) crystal plane slightly shifted to a higher angle and its full width at half maximum decreased as the N_2 flow increased. The electrical resistivity and the emissivity reduced for the plasma annealed AZO/ITO films when the N_2 flow was raised. The optimum H_2/N_2 gas flow was 100/100 for plasma annealed AZO/ITO films in this work for low emissivity application. The emissivity and average visible transmittance for H_2/N_2 = 100/100 plasma annealed AZO/ITO were 0.07 and 80%, respectively, lying in the range of commercially used low emissivity glass.

Keywords: H_2/N_2 plasma; AZO/ITO; low emissivity glass

---

1. Introduction

Low emissivity glass (low-e glass), which owns characteristics of high visible transmittance and infrared reflectance, has been popularly applied in energy-saving architecture [1,2]. Two kinds of low emissivity glass exist: metal-based multilayers and heavily doped wide energy gap semiconductors are reported in the market [3,4]. Silver is the most often used metal for metal-based multilayers. The film thickness of silver must be carefully controlled to possess highly both visible transmittance and infrared reflectance [5]. However, silver is easily oxidized and undergoes poor adhesion with a glass substrate. Some protection and interface layers are usually added on and below the silver films on the glass substrate. Single-layer silver-based multilayer low emissivity glass has at least five layers of films, while double-layer silver ones have more than eight layers [6]. Multilayer film preparation increases production cost. Heavily doped wide energy gap semiconductors which do not need many layers and which may be considered as a substitute for silver-based multilayers applied in low emissivity glass were investigated in this work.

Aluminum-doped zinc oxide (AZO) and tin-doped indium oxide (ITO) are heavily doped wide energy gap semiconductors. Owing to the favorable optical and electrical properties of ITO and AZO [7,8], they are commonly applied in solar cells, organic light-emitting diodes, and energy-saving glass, etc. [9–12]. The Hagen–Rubens relation states emissivity of materials decreases when lowering their electrical resistivity [13]. Recent studies have shown that post treatment could reduce electrical resistivity and increase the visible transmission of ITO and/or AZO films [14–19]. The reduction of electrical resistivity of AZO for H_2 plasma treatment possibly results from desorption of oxygen from the grain boundary or the formation of a complex (such as Zn-H) [14–16]. Muthitamongkol et al. have reported that a decrease in electrical resistivity of AZO films with Ar plasma treatment may be attributed to the conversion of the crystal structure orientation [18]. Wu et al. have reported that a reduction in electrical resistivity for H_2/Ar plasma possibly resulted from reducing negatively-charged oxygens adsorbed on the surface of the grain boundary and providing shallow donor states produced by doped hydrogens [19].

Our research team has fabricated ITO films post-annealed at different H_2/N_2 flows [20]. We have found that optimized H_2/N_2 annealing on ITO can reduce the electrical resistivity of ITO films 58% more than that with pure H_2 annealing. This could be related to the high thermal conductivity of H_2, which is seven times that of N_2 [21]. Plasma energy can be readily dissipated by this high thermal conductivity media during plasma treatment on ITO.

This work investigated the structure, electrical, and optical properties of AZO/ITO films. The AZO/ITO films on glass were prepared by in-line sputtering. Later, the AZO/ITO films were post-annealed in H_2/N_2 plasma at different H_2/N_2 flow ratios (100/0, 100/50, and 100/100). The results indicate that the structure of AZO/ITO films can be modified by H_2/N_2 plasma annealing. The structure and electrical and optical properties of the AZO/ITO films were surveyed. The optimized flow ratio in H_2/N_2 plasma annealing was obtained from the analysis of the measured results.

2. Materials and Methods

The AZO/ITO films were prepared on glass substrates by in-line sputtering. The film thickness of both AZO and ITO was 250 nm. The substrate was not intentionally heated during sputtering. The sputtering target of AZO was ZnO:Al_2O_3 = 98:2 wt.% in composition and 760 × 136 mm^2 in size. The AZO layer was sputtered using a pure Ar flow of 440 sccm, a working pressure of 3 × 10^{-3} Torr, and a sputtering power of 2 kW. The sputtering target of ITO was with In_2O_3:SnO_2 = 90:10 wt.% in composition and 150 × 1500 mm^2 in size. The ITO layer was sputtered under an Ar(85%)/O_2(15%) flow of 40 sccm, a working pressure of 1.6 × 10^{-3} Torr, and a sputtering power of 9kW. The glass substrate was rinsed by ultrasonic treatment in acetone, isopropyl alcohol, and pure water sequentially. After that, the glass was dried with dry N_2.

The AZO/ITO films were plasma annealed with different H_2/N_2 flow ratios of 100/0, 100/50, and 100/100, and were named samples S1, S2, and S3, respectively. Other process parameters used in H_2/N_2 plasma treatment were 25 Torr in gas pressure, 600 W in plasma power and 5 min in process time.

The microstructure and electrical, optical, and emissivity properties of the H_2/N_2 plasma annealed AZO/ITO films were investigated. The crystalline structure and surface morphologies of the AZO/ITO films were explored separately using an X-ray diffractometer (D/MAX-2500 V, Rigaku, Tokyo, Japan) and a scanning electron microscope (SU8000, HITACHI, Tokyo, Japan). Hall measurements (Ecopia HMS-3000, Ecopia, Gyeonggi-do, South Korea) were made in order to deduce the carrier concentration, mobility, and electrical resistivity of the AZO/ITO films. The visible optical transmittance of the AZO/ITO films was surveyed with a UV/VIS/NIR spectrophotometer (PerkinElmer LAMBDA 750, PerkinElmer, Waltham, USA) in the 380~780 nm range. The emissivity of the AZO/ITO films was obtained using an emissivity meter (TSS-5X, Japan Sensor, Tokyo, Japan).

3. Results and Discussions

Figure 1 shows scanning electron micrographs of S1, S2, and S3, respectively. The crystal grains of the AZO/ITO films are shown to grow with increasing N_2 flow.

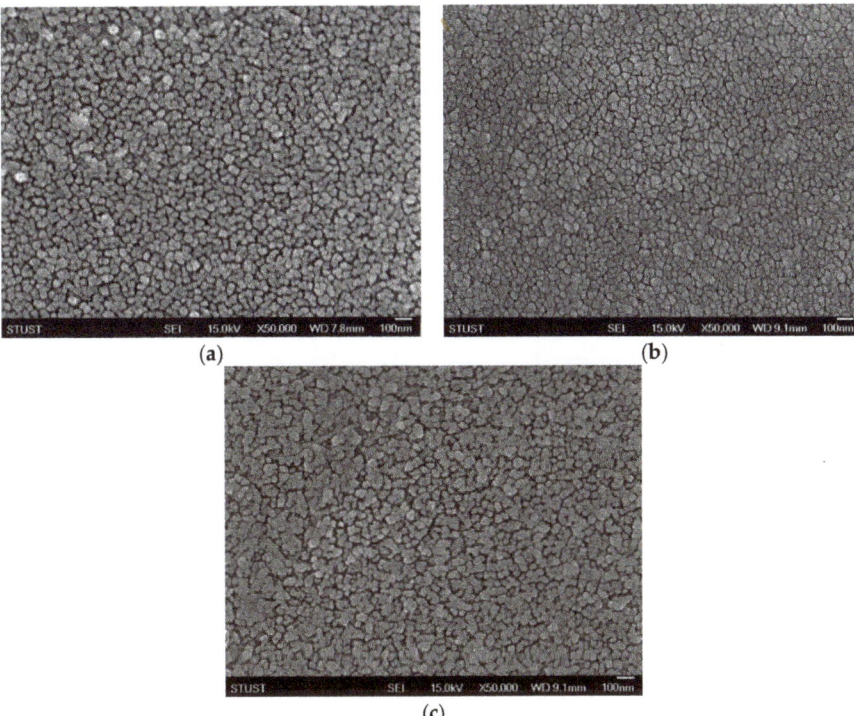

Figure 1. Scanning electron micrographs of the H_2/N_2 plasma processed aluminum-doped zinc oxide (AZO)/tin-doped indium oxide (ITO) films at gas ratios of (**a**) 100/0, (**b**) 100/50, and (**c**) 100/100.

X-ray diffractometer patterns of S1, S2, and S3 are presented in Figure 2, in which two peaks corresponding to the (002) of ZnO and the (222) of ITO may be observed. The peak with respect to (002) of ZnO is seen to slightly shift to a higher angle and its full width at half maximum (FWHM) decreases with increasing N_2 flow.

The (002) peak of ZnO shifting towards a slightly higher angle implies that the distance between the (002) crystal planes of ZnO is decreasing. This phenomenon may be the result of replacing the Zn ions with Al ions [22]. Zinc ions have an ionic radius of 0.74 Å, which is similar to but a little bigger than the radius of Al ions, which is 0.50 Å [23]. The Scherrer equation points out that the grain size of materials increases when the FWMH of the X-ray diffraction peak reduces for the corresponding materials [24]. The FWHM of the (002) peak for ZnO narrows with increasing N_2 flow during H_2/N_2 plasma annealing, as shown in Figure 2, indicating that the grain size of the H_2/N_2 plasma annealed AZO/ITO films increases with N_2 flow during annealing deduced from the Scherrer equation, which is in agreement with the results of the SEM micrographs observed in Figure 1.

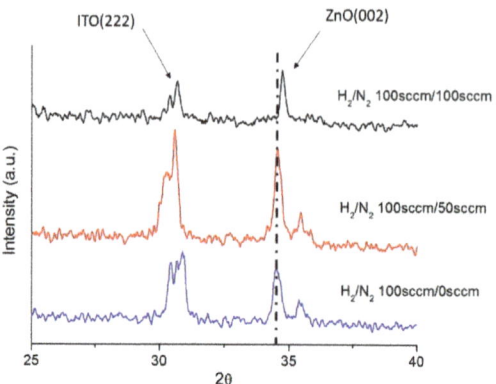

Figure 2. XRD patterns of the H$_2$/N$_2$ plasma processed AZO/ITO films.

The electrical properties of the plasma annealed AZO/ITO films were also investigated. In Figure 3 it is evident that the electrical resistivity of the AZO/ITO films decreases and the carrier concentration of AZO/ITO films increases with N$_2$ flow. Sample S3 had the lowest resistivity, 2.89×10^{-4} Ω-cm, out of S1, S2, and S3. The mobility of S1, S2, and S3 was 19.2, 25.0, and 23.2 cm^2/Vs, respectively.

The reducing electrical resistivity of the AZO/ITO films with increasing N$_2$ flow during the H$_2$/N$_2$ plasma process could result from replacing Zn ions with Al ions and the increase in grain size of the AZO/ITO films. Electrical resistivity of the materials is inversely proportional to the product of carrier concentration and mobility. The XRD diffraction peak corresponding to the (002) ZnO crystal plane slightly shifting to a higher angle, as observed from Figure 2, hints at the replacement of Zn ions with Al ions for the AZO/ITO films. The replacement increases the carrier concentration of the AZO/ITO films, which agrees with the measured results of carrier concentration in Figure 3. The mobility of the materials increases with decreasing defects of the materials. The measured grain size and the carrier concentration of the AZO/ITO films increasing with N$_2$ flow during the H$_2$/N$_2$ plasma process can be observed in Figure 1; Figure 3, respectively. An increase in grain size reduced surface defects and an increase in carrier concentration increased point defects in the AZO/ITO films. This could explain why the mobility of the H$_2$/N$_2$ plasma annealed AZO/ITO films reached a maximum at H$_2$/N$_2$ = 100/50, not at 100/100.

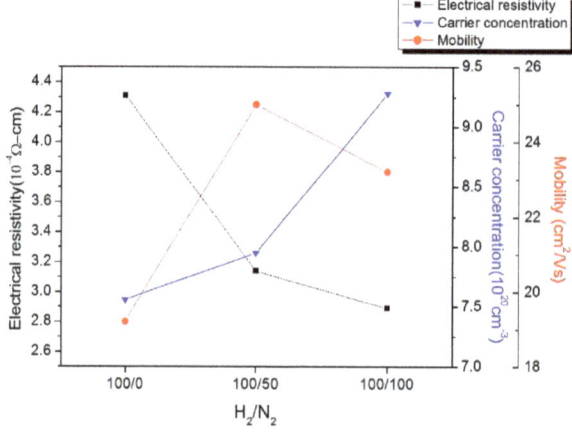

Figure 3. Electrical properties of the H$_2$/N$_2$ plasma processed AZO/ITO films.

Table 1 shows the emissivity and average transmittance in visible light of the H_2/N_2 plasma annealed AZO/ITO films. Figure 4 reveals the transmission spectra of the AZO/ITO films by plasma treatment. The average transmittance in the visible region (380–780 nm) for S1, S2, and S3 was 80%. The emissivity of the AZO/ITO films decreased with increasing N_2 flow. The emissivity of S1, S2, and S3 was 0.10, 0.08, and 0.07, respectively. The emissivity of the materials decreased with their electrical resistivity according to the Hagen–Rubens relation [13]. The electrical resistivity of the AZO/ITO films decreased (Figure 3) and the emissivity of the H_2/N_2 AZO/ITO films also decreased (Table 1) with increasing N_2 flow during the H_2/N_2 plasma process. Our results are in agreement with the Hagen–Rubens relation.

Reports on commercial low-e glass products indicate that they have an emissivity of below 0.45 [25]. Taking into account the obtained results, we deduced that the optimum H_2/N_2 with a flow ratio 100/100 in plasma annealing can produce AZO/ITO films with an emissivity of 0.07.

Table 1. Average transmittance in the visible region (380–780 nm) and emissivity of the H_2/N_2 plasma processed AZO/ITO films.

H_2/N_2 Flow Ratio	100 sccm/0 sccm	100 sccm/50 sccm	100 sccm/100 sccm
Emissivity	0.10	0.08	0.07
Average transmittance in the visible (380–780 nm) region (%)	80%	80%	80%

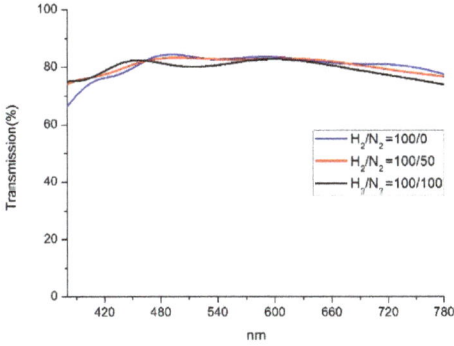

Figure 4. Transmission spectra of the H_2/N_2 plasma with different N_2 flow processed AZO/ITO films.

4. Conclusions

In this study AZO/ITO films were post-annealed in H_2/N_2 plasma at different N_2 flows. Experiment results showed that the N_2 flow in H_2/N_2 plasma annealing of AZO/ITO films affects the microstructure and electrical behavior of AZO/ITO films. The crystal grains grow and the electrical resistivity and emissivity decrease for H_2/N_2 plasma annealed AZO/ITO films as N_2 flow is raised during plasma treatment. The average transmittance in visible wavelengths of all plasma treated AZO/ITO films was about 80%. The decreasing of the electrical resistivity of the AZO/ITO films with increasing N_2 flow during the H_2/N_2 plasma process could be ascribed to the replacement of Zn ions with Al ions and the increase in grain size for the AZO/ITO films. The optimized flow ratio for H_2/N_2 plasma treatment was found to be 100/100. The H_2/N_2 (with a 100/100 flow ratio) plasma annealed AZO/ITO films were able to reach an emissivity of 0.07 and a visible transmittance of 80%, which can be applied to low-e glass.

Author Contributions: S.-C.C. designed and supervised this study and wrote the manuscript with contributions from other co-author. S.-C.C. and H.-T.C. performed the AZO/ITO sample measurement and analysis.

Funding: The authors would like to thank the Ministry of Science and Technology, Taiwan, for their financial support under the grant numbers MOST 105-2221-E-168-010 and MOST 106-2221-E-168-021. This work was partially supported by the Green Energy Technology Research Center.

Conflicts of Interest: The authors declare no conflict of interest.

References

1. Kamalisarvestani, M.; Saidur, R.; Mekhilef, S.; Javadi, F.S. Performance, materials and coating technologies of thermochromic thin films on smart windows. *Renew. Sustain. Energy Rev.* **2013**, *26*, 353–364. [CrossRef]
2. Rezaei, S.D.; Shannigrahi, S.; Ramakrishna, S. A review of conventional, advanced, and smart glazing technologies and materials for improving indoor environment. *Sol. Energy. Mater. Sol. Cells* **2017**, *159*, 26–51. [CrossRef]
3. Karlsson, B.; Valkonen, E.; Karlsson, T.; Ribbing, C.-G. Materials for solar-transmitting heat-reflecting coatings. *Thin Solid Films* **1981**, *86*, 91–98. [CrossRef]
4. Yuste, M.; Galindo, R.E.; Carvalho, S.; Albella, J.M.; Sánchez, O. Improving the visible transmittance of low-e titanium nitride based coatings for solar thermal applications. *Appl. Surf. Sci.* **2011**, *258*, 1784–1788. [CrossRef]
5. Ando, E.; Suzuki, S.; Aomine, N.; Miyazaki, M.; Tada, M. Sputtered silver-based low-emissivity coatings with high moisture durability. *Vacuum* **2000**, *59*, 792–799. [CrossRef]
6. Ding, G.; Clavero, C. Silver-based low-emissivity coating technology for energy-saving window applications. In *Modern Technologies for Creating the Thin-Film Systems and Coatings*, 1st ed.; Nikitenkov, N., Ed.; InTechOpen: London, UK, 2017; pp. 418–422.
7. Yun, J.H.; Kim, J. Double transparent conducting oxide films for photoelectric devices. *Mater. Lett.* **2012**, *70*, 4–6. [CrossRef]
8. Mahmood, K.; Munir, R.; Kang, H.W.; Sung, H.J. An atmospheric pressure-based electrospraying route to fabricate the multi-applications bilayer (AZO/ITO) TCO films. *RSC Adv.* **2013**, *3*, 25741–25751. [CrossRef]
9. Sun, K.; Tang, X.F.; Yang, C.; Jin, D. Preparation and performance of low-emissivity Al-doped ZnO films for energy-saving glass. *Ceram. Int.* **2018**, *44*, 19597–19602. [CrossRef]
10. Kim, H.; Gilmore, C.M.; Horwitz, J.S.; Piqué, A.; Murata, H.; Kushto, G.P.; Schlaf, R.; Kafafi, Z.H.; Chrisey, D.B. Transparent conducting aluminum-doped zinc oxide thin films for organic light-emitting devices. *Appl. Phys. Lett.* **2000**, *76*, 259–261. [CrossRef]
11. Sima, C.; Grigoriu, C.; Antohe, S. Comparison of the dye-sensitized solar cells performances based on transparent conductive ITO and FTO. *Thin Solid Films* **2010**, *519*, 595–597. [CrossRef]
12. Ayachi, B.; Aviles, T.; Vilcot, J.P.; Sion, C. Rapid thermal annealing effect on the spatial resistivity distribution of AZO thin films deposited by pulsed-direct-current sputtering for solar cells applications. *Appl. Surf. Sci.* **2016**, *366*, 53–58. [CrossRef]
13. Hagen, E.; Rubens, H. Über Beziehungen des Reflexions- und Emissionsvermögens der Metalle zu ihrem elektrischen Leitvermögen. *Ann. Phys.* **1903**, *11*, 873–901. [CrossRef]
14. Jiang, Q.J.; Lu, J.G.; Yuan, Y.L.; Sun, L.W.; Wang, X.; Wen, Z.; Ye, Z.Z.; Xiao, D.; Ge, H.Z.; Zhao, Y. Tailoring the morphology, optical and electrical properties of DC-sputtered ZnO:Al films by post thermal and plasma treatments. *Mater. Lett.* **2013**, *106*, 125–128. [CrossRef]
15. Chang, H.P.; Wang, F.H.; Wu, J.Y.; Kung, C.Y.; Liu, H.W. Enhanced conductivity of aluminum doped ZnO films by hydrogen plasma treatment. *Thin Solid Films* **2010**, *518*, 7445–7449. [CrossRef]
16. Cai, P.F.; You, J.B.; Zhang, X.W.; Dong, J.J.; Yang, X.L.; Yin, Z.G.; Chen, N.F. Enhancement of conductivity and transmittance of ZnO films by post hydrogen plasma treatment. *J. Appl. Phys.* **2009**, *105*, 083713. [CrossRef]
17. Lee, J.; Lim, D.; Yang, K.; Choi, W. Influence of different plasma treatments on electrical and optical properties on sputtered AZO and ITO films. *J. Cryst. Growth* **2011**, *326*, 50–57. [CrossRef]
18. Muthitamongkol, P.; Thanachayanont, C.; Samransuksamer, B.; Seawsakul, K.; Horprathum, M.; Eiamchai, P.; Limwichean, S.; Patthanasettakul, V.; Nuntawong, N.; Songsiriritthiguland, P.; et al. The effects of the argon plasma treatments on transparent conductive aluminum-dope zinc oxide thin films prepared by the pulsed DC magnetron sputtering. *Mater. Today Proc.* **2017**, *4*, 6248–6253. [CrossRef]
19. Wu, M.; Huang, T.; Jin, C.; Zhuge, L.; Han, Q.; Wu, X. Effect of Multiple Frequency H_2/Ar Plasma Treatment on the Optical, Electrical, and Structural Properties of AZO Films. *IEEE Trans. Plasma Sci.* **2014**, *42*, 3687–3690. [CrossRef]
20. Chang, S.C. Low pressure H_2/N_2 annealing on indium tin oxide film. *Microelectron. J.* **2007**, *38*, 1220–1225. [CrossRef]

21. O'Hanlon, J.F. *A User's Guide to Vacuum Technology*, 2nd ed.; John Wiley & Sons, Inc.: Hoboken, NJ, USA, 1989; p. 433.
22. Tong, H.; Deng, Z.; Liu, Z.; Huang, C.; Huang, J.; Lan, H.; Wang, C.; Cao, Y. Effects of post-annealing on structural, optical and electrical properties of Al-doped ZnO thin films. *Appl. Surf. Sci.* **2011**, *257*, 4906–4911. [CrossRef]
23. Kittel, C. *Introduction to Solid State Physics*, 8th ed.; John Wiley & Sons, Inc.: Hoboken, NJ, USA, 2004; p. 71.
24. Lu, H.Y.; Chu, S.Y.; Tan, S.S. The characteristics of low-temperature-synthesized ZnS and ZnO nanoparticles. *J. Cryst. Growth* **2004**, *269*, 385–391. [CrossRef]
25. Jelle, B.P.; Kalnæs, S.E.; Gao, T. Low-emissivity materials for building applications: A state-of-the-art review and future research perspectives. *Energy Build.* **2015**, *96*, 329–356. [CrossRef]

© 2019 by the authors. Licensee MDPI, Basel, Switzerland. This article is an open access article distributed under the terms and conditions of the Creative Commons Attribution (CC BY) license (http://creativecommons.org/licenses/by/4.0/).

Article

Research on the High-Performance Electrochemical Energy Storage of a NiO@ZnO (NZO) Hybrid Based on Growth Time

Jiahong Zheng [1,*], Runmei Zhang [1], Kangkang Cheng [1], Ziqi Xu [1], Pengfei Yu [1], Xingang Wang [1] and Shifeng Niu [2]

[1] School of Materials Science and Engineering, Chang'an University, Xi'an 710064, China; zrm199307012222@163.com (R.Z.); 18355098973@163.com (K.C.); xuziqi997@163.com (Z.X.); yupengfei@chd.edu.cn (P.Y.); zyxgwang@chd.edu.cn (X.W.)
[2] College Key Laboratory Automotive Transportation Safety Technology Ministry of Communication, Chang'an University, Xi'an 710064, China; nsf530@163.com
* Correspondence: jhzheng@chd.edu.cn; Tel.: +86-029-8233-7340

Received: 14 December 2018; Accepted: 11 January 2019; Published: 16 January 2019

Abstract: A NiO@ZnO (NZO) hybrid with different reaction times was successfully synthesized by a green hydrothermal method. After comparison, it was found that hydrothermal time had a great impact on specific capacitance. As a supercapacitor electrode of NZO-12h, it exhibited the maximum reversible specific capacitance of 985.0 F/g (3.94 F/cm^2) at 5 mA/cm^2 and 587.5 F/g (2.35 F/cm^2) at 50 mA/cm^2, as well as a high retention of 74.9% capacitance after 1500 cycles at 20 mA/cm^2. Furthermore, the asymmetric electrode device with ZnO-12h and activated carbon (AC) as the positive and negative electrodes was successfully assembled. In addition, the device exhibited a specific capacitance of 85.7 F/g at 0.4 A/g. Moreover, the highest energy density of 27.13 Wh kg^{-1} was obtained at a power density of 321.42 W kg^{-1}. These desirable electrochemical properties demonstrate that the NZO hybrid is a promising electrode material for a supercapacitor.

Keywords: NiO@ZnO; electrochemical performance; supercapacitor

---

1. Introduction

The development of industrialization has caused tremendous pressure on the environment, and it has also seriously affected people's lives. It is necessary to find sustainable and renewable resources [1]. Supercapacitors, excellent energy storage devices, can effectively alleviate the current energy crisis [2–4]. Based on their obvious advantages, such as simple design, high-power density, long cycling lifetime, and short charge/discharge rate [5–7], supercapacitors have attracted much research interest in recent years. However, their development has been limited due to low-energy density, making it difficult to obtain a capacitor with both high-energy density and good power density [8–10]. It is important to find an effective solution to improve the performance of the electrode material.

As an excellent candidate, transition metal oxide can obtain good capacitance characteristics benefiting from its various oxidation states [11]. Nickel oxide (NiO), a low-toxicity and high theoretical specific capacity electrode material, has been widely investigated for energy storage devices [12]. Until now, research on nickel-based materials has made great progress for supercapacitors [13,14]. Gund et al. [14] fabricated micro-belts like β-Ni(OH)$_2$ thin films, and its specific capacitance was 462 F/g at 5 mV/s. As is known, the structure of a material has a direct influence on its performance. To some extent, structural limitations make the conductivity of transition metal oxide unsatisfactory, so that the actual specific capacitance is lower than the theoretical value [15]. It is reported that the electrochemical property of mixed transition metal oxide is superior to single transition metal oxides, and the former can display the synthetic effect of each component. Currently, considerable efforts have been devoted to developing the materials, which exhibit improved electrochemical performance [16–18].

As a single-crystal material, zinc oxide (ZnO) possesses excellent properties, such as wide band gap (3.7 eV), a high-energy density of 650 Ah/g, and excellent chemical stability [19–22]. In addition, the electrical conductivity of ZnO can achieve 230 S/cm [23]. Thus, ZnO can be used as a potential electrode material [24–27]. For example, Hou et al. [27] reported that the ZnO/NiO hierarchical core-shell structure exhibits enhanced pseudocapacitive behaviors, owing to synergistic effect of combining ZnO and NiO/MoO$_2$ composite. Based on this theory, a hybrid (NZO) was prepared by combining NiO and ZnO in our work. The flower-like NZO electrode material has a large surface area, allowing the electrolyte solution to have more contact with the active material. Thus, NZO electrode material has enhancing electrochemical performance.

In this study, NZO on Ni foam with different reaction times was fabricated using a simple and fast hydrothermal method. Their structure and electrochemical properties were also investigated successfully. It was found that NZO-12h exhibits the best electrochemical performance compared with other electrode materials. Further, it shows a high specific capacitance of 3.94 F/cm^2 at 5 mA/cm^2 and high rate capability of 59.6% due to its synergistic properties, growth on conductive substrate, and unique flower-like structure.

2. Materials and Methods

All reagents required for the experiment were of analytical grade, and purchased from Sinopharm Chemical Reagent Co., Ltd. (Shanghai, China).

The NZO was prepared using a facile hydrothermal method. First, Ni foam (2 cm × 2 cm) was pretreated with 5% HCl solution to remove the oxide layer on its surface, and then it was ultrasonically washed with acetone, ethanol, and deionized (DI) water for the same time, with each step sustained for 15 min. Then, 2 mmol nickel nitrate, 4 mmol zinc nitrate, and 24 mmol urea were dispersed into 50 mL DI water with continuous stirring for 30 min to form a green transparent solution. The solution was transferred into a 100 mL autoclave and a piece of pre-treated Ni foam immersed. The autoclave was sealed and heated at 373 K for 12 h to carry out the hydrothermal reaction, and then naturally cooled to room temperature. Next, the Ni foam was rinsed with DI water and ethanol several times, and dried it at 333 K for 5 h. Finally, cleaned Ni foam was put in the tube furnace at 523 K for 2 h, heating rate 5 °C/min, and the product was labeled as NZO-12h. In order to study the effect of hydrothermal time on the properties of electrode materials, analogous methods were used to fabricate other NZO samples with different reaction times of 3 h, 6 h, and 24 h. These samples were denoted as NZO-3h, NZO-6h, and NZO-24h, respectively. Schematic illustration of synthesis process for NZO-12h is shown in Figure 1.

As-prepared products were analyzed using X-ray diffraction (XRD) measurements and recorded with a Rigaku D/max 2500PC diffractometer with Cu Ka radiation (λ = 0.154156 nm). X-ray photoelectron spectroscopy (XPS) was taken through an ESCA-LAB Mk II (Vacuum Generators) spectrometer (Thermo Electron Corporation, New York, America). The morphology and microstructure of the as-prepared sample were characterized by field emission scanning electron microscope (SEM, S-4800, Hitachi, Tokyo, Japan) and transmission electron microscope (TEM, Tecnai F30G2, FEI, Oregon, America) techniques, and chemical composition clearly recorded by energy dispersive X-ray spectrometer (EDS) tests. In order to ensure the quality of the picture, the sample was sputtered with a thin Au-Pt. Surface area analysis was conducted on an ASAP2020HD88 instrument (Micromeritics Instrument Corporation, Shanghai, China) based on the Brunauer–Emmett–Teller (BET) theory. Cyclic voltammetry (CV), galvanostatic charge-discharge (GCD) curves, and electrochemical impedance spectroscopy (EIS) measurements were examined using a three-electrode system on CHI 660E electrochemical station (Chenhua, Shanghai, China) in 6 M KOH. The sample was used as the working electrode, a platinum foil (2 cm × 2 cm), and a saturated calomel electrode as the counter and the reference electrode, respectively. The mass loading of the active material on the Ni foam was about 4 mg, which was determined by weighing the Ni foam substrate before and after applying the active material. And the specific capacitance of the single electrode was calculated using the equation:

$$C = i\Delta t / (S\Delta V) \qquad (1)$$

$$C = i\Delta t / (m\Delta V) \qquad (2)$$

where C is the specific capacitance, i is the discharge current, Δt is the discharge time, S and m are assigned as the area and mass of the electrode material, and ΔV represents the potential window.

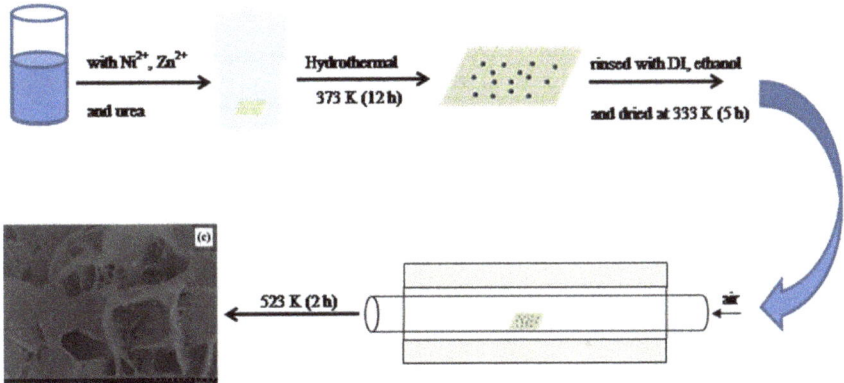

Figure 1. Schematic illustration of synthesis process for NZO-12h.

The electrochemical performance of asymmetric supercapacitor was examined in a two-electrode system. The NZO-12h sample was used as the positive electrode and activated carbon as the negative electrode, which was separated by a piece non-woven fabric. Further, a negative electrode was prepared using activated carbon and polytetrafluoroethylene (PTFE) with mass ratio of 9:1. After, homogeneous slurry was formed through continuous mixing, and then obtained slurry was spread on cleaned Ni foam. Next, the Ni foam was dried at 340 K for 6 h in a vacuum oven.

The mass loading of positive and negative electrode was calculated according to the following equation:

$$\frac{m_+}{m_-} = \frac{C_- \times \Delta E_-}{C_+ \times \Delta E_+} \qquad (3)$$

where m (g) represents the mass loading, C is specific capacitance, and ΔE is potential window, respectively, of positive (+) and negative (−) electrodes.

3. Results

The XRD measurement was used to analyze phase information of the NZO-12h electrode, and the result is displayed in Figure 2. Three strong peaks can be found at 44.51°, 51.84°, and 76.41°, which corresponds to (110), (200), and (220) diffraction of Ni (JCPDS 65-2865). There were two diffraction peaks which matched well with the (100) and (101) planes of ZnO (JCPDS 36-1451). The characteristic peak at 43.01° corresponds to the (200) planes of NiO (JCPDS 47-1049). The XRD pattern indicates two phases of NiO and ZnO could be indexed in the hybrid. Therefore, NZO was successfully fabricated using the hydrothermal method.

The chemical composition and oxidation state of the samples were evaluated by XPS measurements. The fitted spectra were obtained by Gaussian simulation method, as shown in Figure 3. The Ni $2p_{3/2}$ main peak was located at 855.4 eV and its satellite peak at 861.1 eV, while the Ni $2p_{1/2}$

main peak was centered at 873.8 eV and its satellite peak at 879.9 eV, respectively (Figure 3a). The result, which is consistent with the previously reported values, further confirms that NiO can be detected in as-prepared samples [28]. In the Figure 3b, it is clear that the prominent peaks of Zn 2p$_{3/2}$ and Zn 2p$_{1/2}$ appear at 1021.2 eV and 1044.5 eV, which was attributed to Zn^{2+} in the sample [29]. In the O 1s region, there are three peaks, as shown in Figure 3c. Two peaks can be found at 527.9 eV and 530.5 eV, which represents O^{2-} (Ni) and O^{2-} (Zn), respectively. In addition, a pronounced peak was observed at 532.8 eV, which corresponds to H_2O [28]. These results clearly determine the formation of ZnO and NiO in this sample.

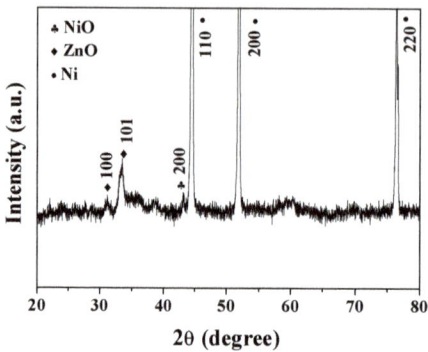

Figure 2. XRD pattern of NZO-12h on Ni foam.

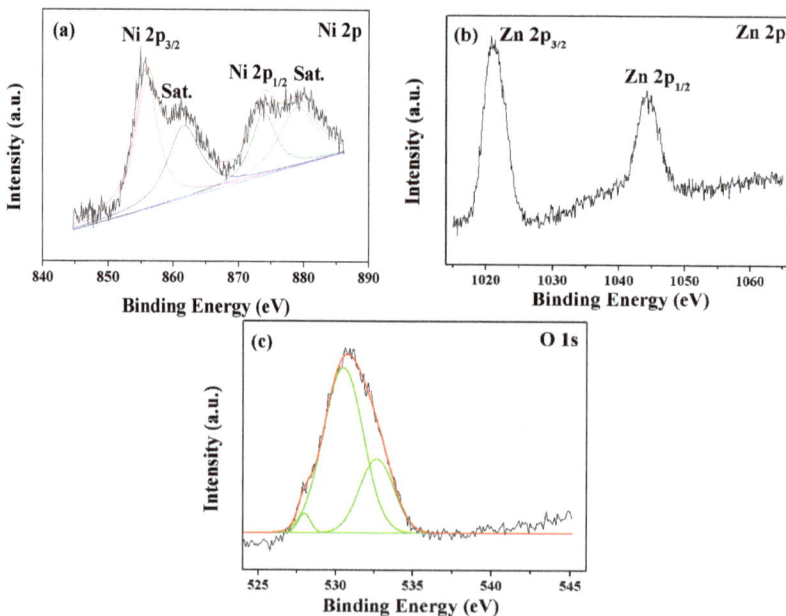

Figure 3. X-ray photoelectron spectroscopy (XPS) survey spectra of NZO-12h: (**a**) Ni 2p; (**b**) Zn 2p; (**c**) O 1s.

Figure 4 exhibits the SEM images at different magnifications of NZO-3h (Figure 4a–c), NZO-6h (Figure 4d–f), NZO-12h (Figure 4g–i), and NZO-24h (Figure 4j–l), respectively. From low-magnification SEM images (Figure 4a,d,g,j), it can be found that NZO materials grow on the Ni foam, and present the corresponding structure and its distribution states. For the NZO-3h sample, it exhibits a nanosheet structure with an average diameter of 720 nm. Besides, the edge of these nanosheets has a jagged structure, which can be observed from a high magnification of Figure 4b,c. For NZO-6h, the as-prepared sample was composed of nanosheets with an average diameter of 540 nm, and these nanosheets form a flower-like structure. Notably, the center of the flower-like structure has a large accumulation of nanosheets, which affects the electrochemical conductivity of the electrode material. For the NZO-12h sample, it can be seen that a 3D sphere was made up of intertwined nanosheets, and the nanosheets were supported by each other, forming a flower-like structure. The diameter of nanosheets was about 610 nm. Moreover, there were a few pores in these nanosheets, as shown in Figure 4i, which facilitates the diffusion of electroactive species. Such unique porous structure provides numerous pathways for effective active sites electron/ion transport, which also accelerates the occurrence of redox reactions. Thus, the NZO-12h sample has excellent electrochemical performance. Obviously, when the reaction time is 24 h, the structure of the sample is still close to the flower-like; however, it is different from other electrode materials. The magnified image (Figure 4l) illustrates NZO-24h was composed of many short-rods and nanosheets.

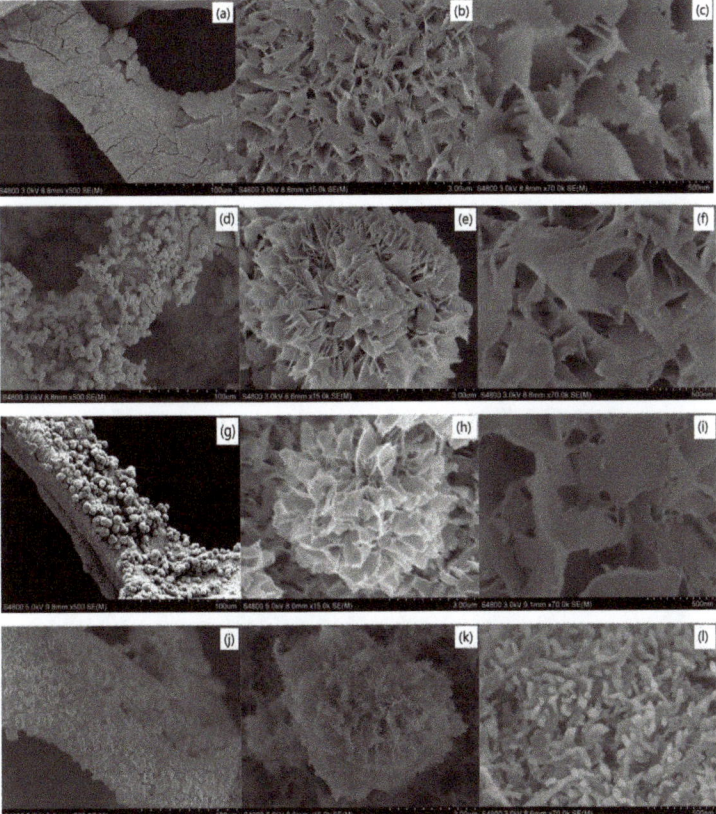

Figure 4. FESEM images of (**a–c**) NZO-3h; (**d–f**) NZO-6h; (**g–i**) NZO-12h; and (**j–l**) NZO-24h.

EDS is an important technique to examine and analyze elemental component of samples. EDS was performed at a working distance of 15 mm and an acceleration voltage of 15 kV. The typical EDS spectrum of the NZO-12h sample is shown in Figure 5. It can be clearly seen that Ni, O, and Zn were detected in the as-obtained sample, which is consistent with the XPS result.

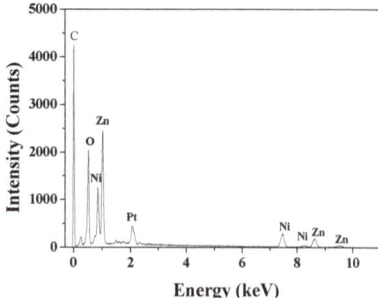

Figure 5. Typical EDS spectrum of the NZO-12h.

To observe the detailed structural characteristic of the NZO-12h sample, TEM and high-resolution-TEM (HRTEM) were carried out, as shown in Figure 6. Further, TEM can be used to distinguish between crystalline and amorphous structures. Meanwhile, precise information about the surface morphology can be provided. The low-magnification TEM image is exhibited in Figure 6a, which reveals the existence of interdigitated nanosheets for NZO-12h. The result coincided with the corresponding SEM patterns. The HRTEM of the NZO-12h sample is displayed in Figure 6b–d, and the images show that there are two kinds of diffraction fringes with the lattice spacing of about 0.237 nm and 0.241 nm, which can be ascribed to the (111) planes of NiO and (101) planes of ZnO. The diffraction spots in the selected area electron diffraction (SAED) pattern suggest crystallization property of the as-prepared sample is unsatisfactory. Therefore, the conclusion could be made undoubtedly that the flower-like NZO sample has been successfully prepared using the hydrothermal method.

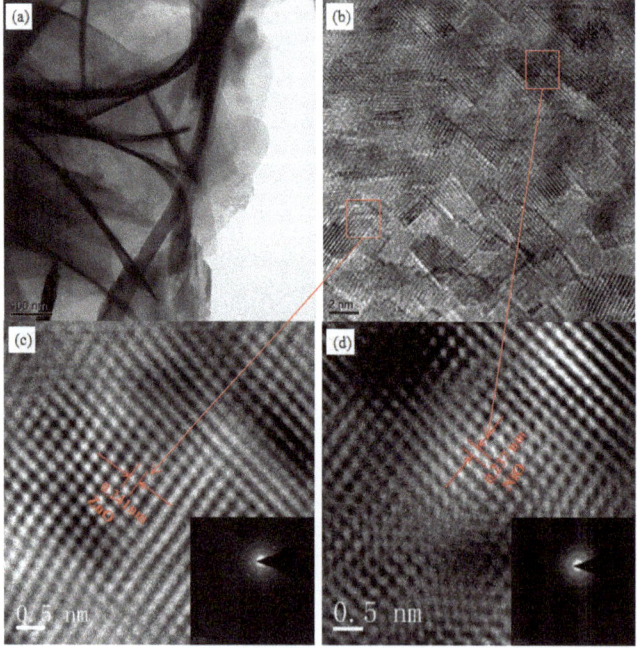

Figure 6. TEM characterization of NZO-12h: (**a**) TEM image; (**b–d**) high-resolution-TEM images; inset in (**c,d**) are corresponding selected area electron diffraction (SAED) images of ZnO and NiO.

The nitrogen adsorption and desorption test was used to analyze the porous structure, and the result is displayed in Figure 7. The flower-like NZO-12h sample exhibits a high Brunauer–Emmett–Teller (BET) surface area of 20.3350 cm^2/g, which is beneficial for the transport and diffusion of electrolyte. The pore size distribution of NZO-12h is shown in Figure 7b, and an average pore diameter of 2 nm was obtained by the Barret-Joyner-Halenda (BJH) method using the adsorption branch of the isotherm. From this feature, it can be concluded that the NZO-12h sample has micropores. Consequently, the NZO-12h sample has good electrochemical performance.

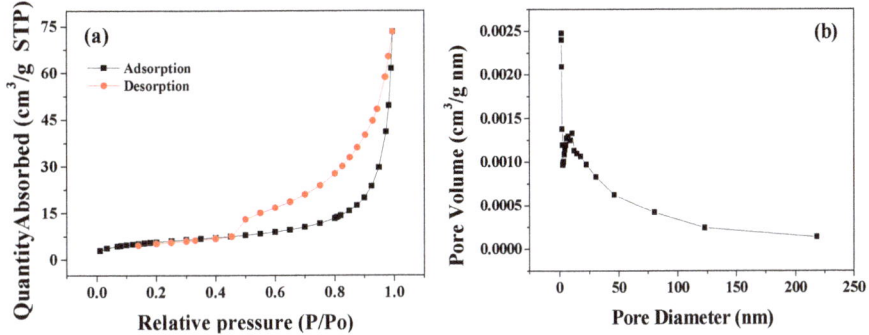

Figure 7. Nitrogen adsorption and desorption isotherms and pore size distribution of NZO-12h.

To explore the effect of structure on performance, a series of electrochemical tests of NZO materials were performed. The CV curves of NZO electrodes (3 h, 6 h, 12 h, and 24 h) were measured at 5 mV/s within potential windows of 0–0.5 V, and the results are presented in Figure 8a. Obviously, well-defined anodic and cathodic peaks of each curve can be observed. The special shape reflects the pseudocapacitive characteristics of the NZO, which is different from the CV curves of the double-layer capacitor [30]. Clearly, it can be found NZO-12h exhibits the largest enclosed area, implying its specific capacitance is better than that of NZO-3h, NZO-6h, and NZO-24h electrodes. As is known, electrochemical performance is related to the structure of samples. From these images it clearly shows that the morphology of NZO-24h is different from NZO-12h, which consists of short-rods and sheets. However, intertwined nanosheets of flower-like NZO-12h create a number of pores that have a large effect on increasing the capacitance of the electrode material. In Figure 8a, NZO-3h and NZO-6h samples exhibit almost the same double layer electrochemical surface area. The enclosed area is smaller when the reaction time is 6 h, which is caused by poor electrical conductivity. Inconsistent redox peak position change trend is mainly caused by the difference of pore structure. Figure 8b describes the CV curves of the NZO-12h at different scan rates (5, 10, 25, and 50 mV/s). As the scan rates increases, the distance between redox peaks becomes wide due to polarization. Moreover, the CV shape is similar at different scan rates, which means that the electrode material has good reversibility [31], as shown in the inset of Figure 8b. Therefore, it is necessary to further investigate the electrochemical properties of the NZO-12h.

The GCD tests of different NZO electrodes were conducted at a current density of 5 mA/cm^2, as shown in Figure 9a. The area specific capacitances of the NZO-3h, NZO-6h, NZO-12h, and NZO-24h were 1.49 F/cm^2, 0.41 F/cm^2, 3.94 F/cm^2, and 3.32 F/cm^2. Meanwhile, the mass specific capacitances of these materials were measured to be 372.5 F/g, 102.5 F/g, 985.0 F/g, and 830.0 F/g. It can be clearly found that the NZO-12h electrode has the highest specific capacitance, which is in agreement with the CV result. The electrochemical performance of NZO-12h is better than NZO-3h, 6h, and 24h, which was mainly attributed to its unique structural characteristics. Further, when the hydrothermal reaction is 12 h, more active material grows on Ni foam. And its approximate flower-like structure also promotes the contact between sample and electrolyte, thereby exhibiting better performance.

The relationship between specific capacitances and hydrothermal time is shown in Figure 9b. It can be found the NZO-12h exhibits higher specific capacitance than others.

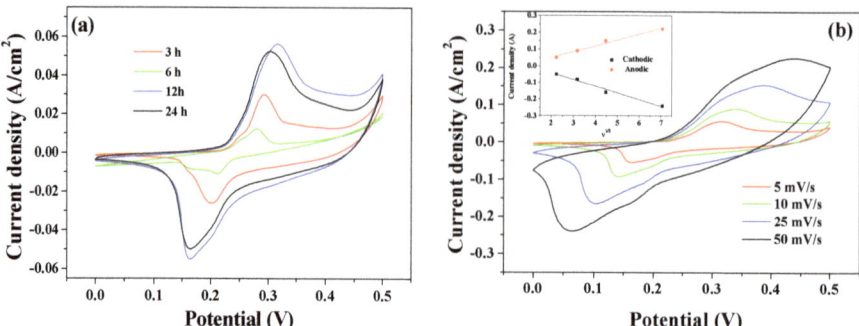

Figure 8. (a) CV curves of NZO electrodes prepared using different hydrothermal times at a scan rate of 5 mV/s; (b) rate performance curves of the NZO-12 h electrode, the insets show the i_p vs. V plots of the corresponding CV curves.

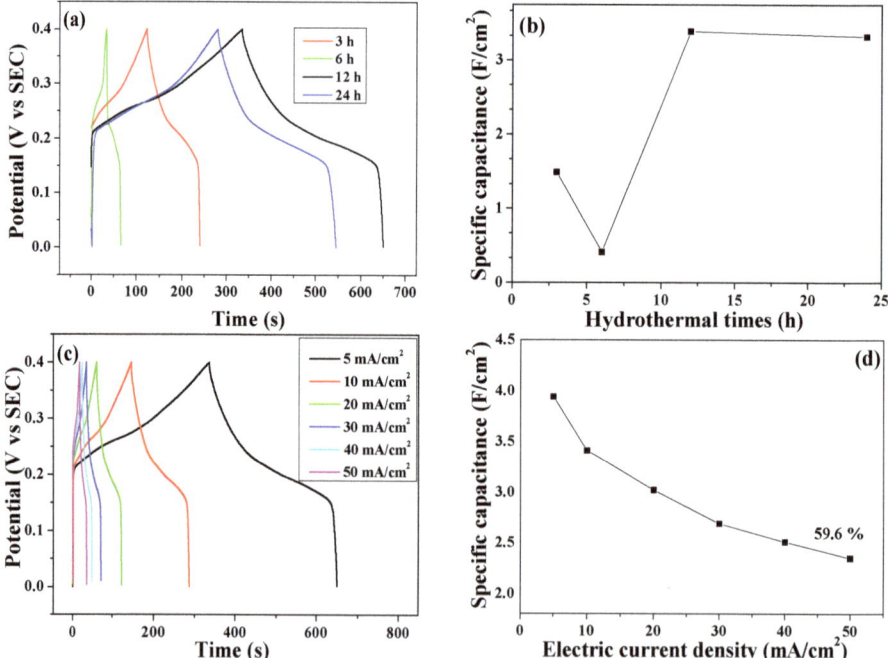

Figure 9. (a) GCD curves at a current density of 5 mA/cm² for NZO electrodes with different hydrothermal times; (b) the specific capacitances of the NZO with different hydrothermal times; (c) GCD curves of the NZO-12h electrode at different current densities; (d) rate capacitance calculation of the NZO-12h electrode.

Figure 9c shows the GCD curves of the NZO-12h electrode at different current densities. Clearly, there exists a discharge platform for each non-linear GCD curve. Actually, redox reaction is commonly associated with pseudocapacitive charge/discharge process. When the current density is 5 mA/cm², the GCD curve has the longest discharge time. Rate capacitance calculation of NZO-12h derived from Figure 9c is shown in Figure 9d, and area specific capacitances are 3.94, 3.41, 3.02, 2.69, 2.51, and 2.35 F/cm² at the current density of 5, 10, 20, 30, 40, and 50 mA/cm², respectively. According to Equation (2), the mass specific capacitances of 985.0 F/g, 852.5 F/g, 755 F/g, 673.1 F/g, 627.5 F/g, and 587.5 F/g were obtained at the same current density, respectively. In addition, NZO-12h maintains 59.6% of the maximum capacitance at 50 mA/cm², highlighting its excellent rate capability. As expected, NZO-12h has better electrochemical performance than previous reports. In detail, specific capacitances of different electrode materials are shown in Table 1. The high specific capacitance was attributed to the unique structure of NZO-12h. Further, the large specific surface area of such a flower-like structure allows the electrode material to have more active sites, which is significant to the performance of itself. Thus, the flower-like NZO-12h has the potential to meet the demand of supercapacitor applications.

Table 1. Specific capacitances of different electrode materials.

Material	Structure	Specific Capacitance	Reference
NiO@PC	Hierarchical structure	57 mF/cm² at 5 mA/cm²	[32]
NiO@C@Cu₂O hybrid	Core-shell heterostructure	2.18 F/cm² at 1 mA/cm²	[33]
HNCS-NiO	Hierarchical structure	880.6 mF/cm² at 0.8 mA/cm²	[34]
NiO/MnO₂	Heterostructure	286 mF/cm² at 0.5 mA/cm²	[35]
ZnO/MnO₂@carbon cloth	Core-shell structure	138.7 mF/cm² at 1 mA/cm²	[36]
NiO/Ni(OH)₂/PEDOT	Nanoflower structure	404.1 mF/cm² at 4 mA/cm²	[37]
NZO-12h	Flower-like structure	3.94 F/cm² at 5 mA/cm²	Our work

Long stability is an important parameter to determine whether the as-prepared material can be effectively applied in supercapacitors, and the result is shown in Figure 10. After 1500 cycles, 74.9% capacitance remained, indicating its high cycling stability. The GCD curves for first seven and last seven cycles are shown in Figure 10b,c. It can be found that the shape of the curves is similar, meaning that there is no significant structural change.

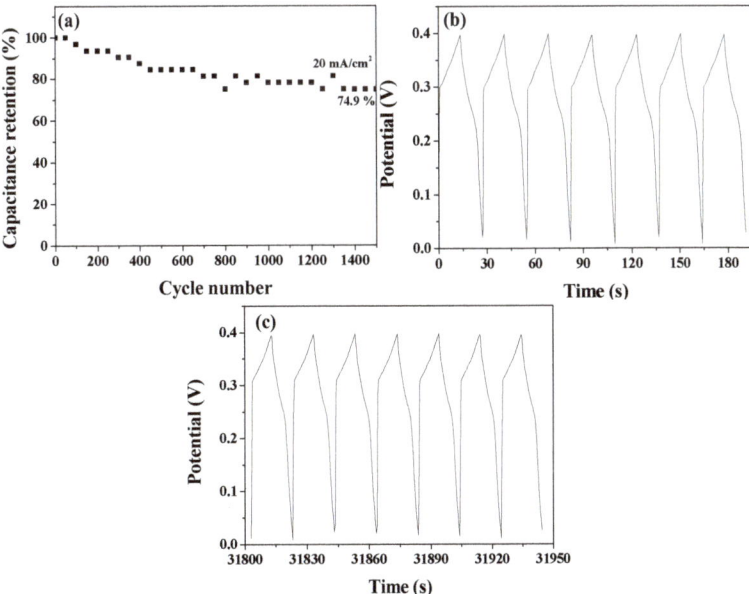

Figure 10. (a) 1500 cycle performance of the NZO-12h at a current density of 20 mA/cm²; (b) first 7 cycles of NZO-12h; (c) final 7 cycles of NZO-12h.

To further understand the electrochemical behavior, the EIS plots of the NZO with different hydrothermal times were collected between 0.01 Hz and 10 kHz, as shown in Figure 11a. Generally speaking, the EIS curve includes a linear portion of the low-frequency region and a non-linear portion of the high-frequency region. For the electrode material, its stability properties, conductivity, and charge–discharge rate [38] are better when the equivalent series resistance (R_s) value is smaller. The diameter of a semicircle at high-frequency represents charge–transfer resistance R_{ct}. The slope of the line can be ascribed to the diffusion and kinetics process called the Warburg impedance (Z_w). Importantly, a steep line indicates the better capacitive behavior in low-frequency region. Thus, NZO-6h is unable to be a promising electrode material. However, the R_s values of the three samples (NZO-3h, NZO-12h, and NZO-24h) are particularly close, being 0.35 Ω, 0.31 Ω, and 0.33 Ω. R_{ct} values are 3.39 Ω, 1.25 Ω, and 1.74 Ω, respectively. It can be concluded that NZO-12h has the smallest R_s and R_{ct}. Moreover, the slope of the NZO-12h is closer to the vertical line, representing a more rapid ion-diffusion-transfer rate. Therefore, NZO-12h is an excellent candidate for the supercapacitor, which is consistent with above analysis.

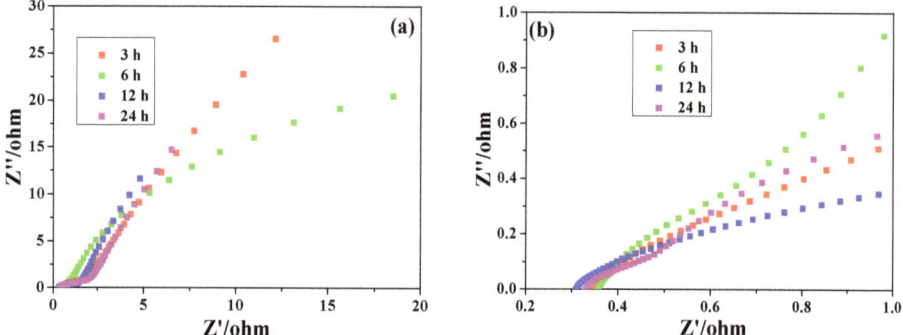

Figure 11. (a) EIS plots of NZO electrodes with different hydrothermal time; (b) the corresponding high-frequency region of EIS.

In order to explore practical applications of the flower-like NZO-12h, it is necessary to fabricate an asymmetric capacitor. The NZO-12h was regarded as a positive electrode and AC as a negative electrode, and the electrochemical performance was tested in 6 M KOH electrolyte. After calculation, the mass ratio of the NZO-12h and AC was 1:3.98. Figure 12a shows the CV curves of AC at different scan rates in the potential window of −1.0–0 V. It was found that the shape of the curve was close to a rectangle without redox peak, indicating the behavior of an electrical double-layer capacitance. When the scan rate increases from 5 to 50 mV/s, the capacitance of the AC remains 73.4%, indicating its good rate capability. Figure 12b shows the CV curves of AC and NZO-12h, which was recorded in a three-electrode system and tested at a scan rate of 10 mV/s. There is no doubt that the positive and negative electrodes were well matched each other for assembling asymmetric supercapacitors. Figure 12c exhibits the CV curves of the asymmetric supercapacitor at different scan rates of 5, 10, 25, and 50 mV/s with the potential of 0–1.5 V. It was found that the shape of the CV curves were nearly quasi-rectangular, even at a scan rate of 50 mV/s, indicating the asymmetric capacitor device has low resistance. The GCD curves of the asymmetric supercapacitor at different current densities were shown in Figure 12d. According to GCD measurement, energy density (E) and power density (P) can be calculated as follows [39]:

$$E = \frac{1}{2 \times 3.6} CV^2 \qquad (4)$$

$$P = E \times 3600/\Delta t \qquad (5)$$

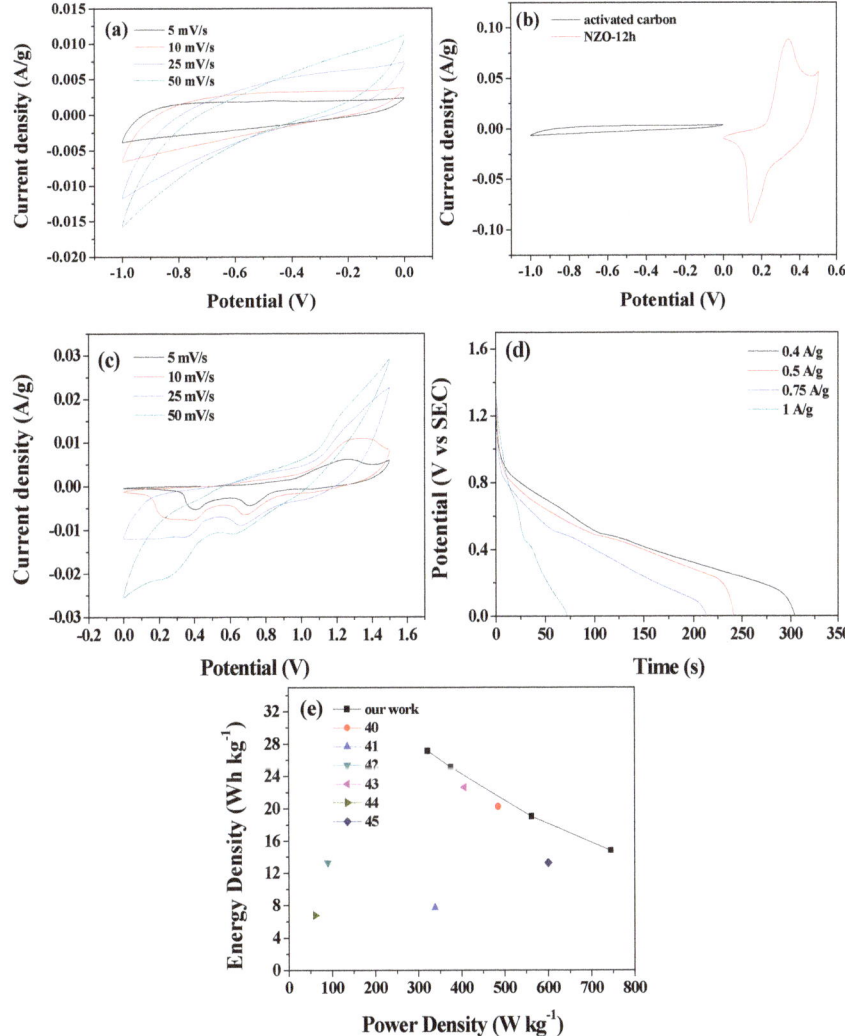

Figure 12. (**a**) CV curve of AC at different scan rates; (**b**) CV curve of AC and NZO-12h separately tested in a three-electrode system at 10 mV/s; (**c**) CV curve of asymmetric supercapacitor at different scan rates; (**d**) GCD curve of asymmetric supercapacitor at different current density; (**e**) Ragone plots of asymmetric supercapacitor.

In this equation, E (Wh kg^{-1}) and P (W kg^{-1}) are the energy density and power density. C, V, and Δt are specific capacitance, operating voltage, and discharge time, respectively.

Ragone plot relative to the corresponding energy and power densities of the asymmetric supercapacitor was shown in Figure 12e. Clearly, the highest energy density of 27.13 Wh kg^{-1} was obtained at a power density of 321.42 W kg^{-1}, and still remains 14.81 Wh kg^{-1} at a high-power density of 744.72 W kg^{-1}. The obtained values are superior to results found in the literature of other asymmetric systems, such as fabricated CNFs/MnO$_2$//AC (20.3 Wh kg^{-1} at 485 W kg^{-1}) [40], MnO$_2$//AC (7.8 Wh kg^{-1} at 338 W kg^{-1}) [41], the novel CNT@NCT@MnO$_2$ composites (13.3 Wh kg^{-1} at 90 W kg^{-1}) [42], FeCo$_2$O$_4$@MnO$_2$ nanostructures on carbon fibers (22.68 Wh kg^{-1} at 406.01 W kg^{-1}) [43], graphene-MnO$_2$//graphene-MnO$_2$ (6.8 Wh kg^{-1}, 62.0 W kg^{-1}) [44], and AC//MnO$_2$-CNTs (13.3 Wh kg^{-1} at 600.0 W kg^{-1}) [45].

4. Conclusions

In summary, NZO with various hydrothermal time was fabricated through a facile, simple, and low-cost hydrothermal method. For four kinds of electrode materials, the NZO-12h exhibited better electrochemical performance than the others (NZO-3h, NZO-6h, and NZO-24h). This phenomenon was attributed to the large specific surface area of the NZO-12h sample, which is advantageous for the occurrence of redox reactions. Electrochemical results show the NZO-12h electrode exhibited a high specific capacitance of 985.0 F/g (3.94 F/cm^2) at 5 mA/cm^2 and 587.5 F/g (2.35 F/cm^2) at 50 mA/cm^2. Besides, 74.9% capacitance retention was obtained after 1500 cycles charge/discharge cycles, which indicates excellent cycling stability of NZO-12h. Moreover, the electrochemical performance of an asymmetric supercapacitor was measured, and the specific capacitance of the device was 85.7 F/g at 0.4 A/g. Moreover, the device can deliver a maximum energy density of 27.13 Wh kg^{-1} at a power density of 321.42 W kg^{-1}. Therefore, the obtained NZO-12h hybrid is a promising candidate for asupercapacitor. Importantly, our present study provides a strategy for the preparation of other hybrid metal oxides.

Author Contributions: S.N. conceived and designed the experiments; Z.X. fabricated and R.Z. characterized the sample; K.C. and Z.X. collaborated in XRD and SEM measurement; J.Z., P.Y., and X.W. analyzed the data. All authors discussed the experiment results and contributed to writing the paper.

Funding: This research was funded by National Natural Science Foundation of China (Grant Nos. 21607013, 51602026), the Special Fund for Basic Scientific Research of Central Colleges of Chang'an University (NOs. 300102318108, 300102228203, 300102318402, 300102318106), the Fund Project of Shaanxi Key Laboratory of Land Consolidation (2018-JC01), and the College Students' Innovation and Entrepreneurship Project (201810710127).

Conflicts of Interest: The authors declare no conflict of interest.

References

1. Cai, G.F.; Wang, X.; Cui, M.Q.; Darmawan, P.; Wang, J.X.; Eh, A.L.S.; Lee, P.S. Electrochromo-supercapacitor based on direct growth of NiO nanoparticles. *Nano Energy* **2014**, *12*, 258–267. [CrossRef]
2. Zhang, X.; He, B.L.; Zhao, Y.Y.; Tang, Q.W. A porous ceramic membrane tailored high-temperature supercapacitor. *J. Power Sources* **2018**, *379*, 60–67. [CrossRef]
3. Dubala, D.P.; Gomez-Romero, P.; Sankapald, B.R.; Holze, R. Nickel cobaltite as an emerging material for supercapacitors: An overview. *Nano Energy* **2015**, *11*, 377–399. [CrossRef]
4. Dubal, D.P.; Chodankar, N.R.; Kim, D.H.; Gomez-Romero, P. Towards flexible solid-state supercapacitors for smart and wearable electronics. *Energy Environ. Sci.* **2018**, *4*, 2065–2129. [CrossRef]
5. Kumar, A.; Sanger, A.; Kumar, A.; Chandra, R. Single-step growth of pyramidally textured NiO nanostructures with improved supercapacitive properties. *Int. J. Hydrog. Energy* **2017**, *42*, 6080–6087. [CrossRef]
6. Emamdoust, A.; Shayesteh, S.F. Surface and electrochemical properties of flower-like Cu-NiO compounds. *J. Alloy. Compd.* **2018**, *738*, 432–439. [CrossRef]
7. Li, Y.; Li, Z.; Shen, P.K. Simultaneous Formation of Ultrahigh Surface Area and Three-Dimensional Hierarchical Porous Graphene-Like Networks for Fast and Highly Stable Supercapacitors. *Adv. Mater.* **2013**, *25*, 2474–2480. [CrossRef] [PubMed]
8. Augustyn, V.; Simon, P.; Dunn, B. Pseudocapacitive oxide materials for high-rate electrochemical energy storage. *Energy Environ. Sci.* **2014**, *7*, 1597–1614. [CrossRef]
9. An, L.; Xu, K.B.; Li, W.Y.; Liu, Q.; Li, B.; Zou, R.J.; Chen, Z.; Hu, J.Q. Exceptional pseudocapacitive properties of hierarchical NiO ultrafine nanowires grown on mesoporous NiO nanosheets. *J. Mater. Chem. A* **2014**, *2*, 12799–12804. [CrossRef]
10. Wang, Q.; Yan, J.; Fan, Z.J. Carbon materials for high volumetric performance supercapacitors: design, progress, challenges and opportunities. *Energy Environ. Sci.* **2016**, *9*, 729–762. [CrossRef]
11. Huang, M.; Li, F.; Zhang, Y.X.; Li, B.; Gao, X. Hierarchical NiO nanoflake coated CuO flower core-shell nanostructures for supercapacitor. *Ceram. Int.* **2014**, *40*, 5533–5538. [CrossRef]

12. Yadav, A.A.; Chavan, U.J. Electrochemical supercapacitive performance of spray-deposited NiO electrodes. *J. Electron. Mater.* **2018**, *47*, 3770–3778. [CrossRef]
13. Dubal, D.P.; Fulari, V.J.; Lokhande, C.D. Effect of morphology on supercapacitive properties of chemically grown β-Ni(OH)$_2$ thin films. *Microporous Mesoporous Mater.* **2012**, *151*, 511–516. [CrossRef]
14. Gund, G.S.; Dubal, D.P.; Shinde, S.S.; Lokhande, C.D. One step hydrothermal synthesis of micro-belts like β-Ni(OH)$_2$ thin films for supercapacitors. *Ceram. Int.* **2013**, *39*, 7255–7261. [CrossRef]
15. Yan, Z.; Guo, C.; Yang, F.; Zhang, C.C.; Mao, Y.Q.; Cui, S.X.; Wei, Y.H.; Xu, L.C. Cliff-like NiO/Ni$_3$S$_2$ directly grown on Ni foam for battery-type electrode with high area capacity and long cycle stability. *Electrochim. Acta* **2017**, *251*, 235–243. [CrossRef]
16. Dubal, D.P.; Holze, R.; Gomezromero, P. Development of hybrid materials based on sponge supported reduced graphene oxide and transition metal hydroxides for hybrid energy storage devices. *Sci. Rep.* **2014**, *4*, 7349. [CrossRef] [PubMed]
17. Wolfart, F.; Dubal, D.P.; Vidotti, M.; Gómezromero, P. Hybrid core-shell nanostructured electrodes made of polypyrrole nanotubes coated with Ni(OH)$_2$ nanoflakes for high energy-density supercapacitors. *RSC Adv.* **2016**, *6*, 15062–15070. [CrossRef]
18. Cao, M.X.; Yang, S.W.; Zhang, Y.T.; Song, X.X.; Che, Y.L.; Zhang, H.T.; Yu, Y.; Ding, G.Q.; Zhang, G.Z.; Yao, J.Q. Tunable amplified spontaneous emission in graphene quantum dots doped cholesteric liquid crystals. *Nanotechnology* **2017**, *28*, 245402. [CrossRef] [PubMed]
19. Favero, V.O.; Oliveira, D.A.; Lutkenhaus, J.L.; Siqueira, J.R. Layer-by-layer nanostructured supercapacitor electrodes consisting of ZnO nanoparticles and multi-walled carbon nanotubes. *J. Mater. Sci.* **2018**, *53*, 6719–6728. [CrossRef]
20. Lu, F.; Cai, W.; Zhang, Y. ZnO hierarchical micro/nanoarchitectures: solvothermal synthesis and structurally enhanced photocatalytic performance. *Adv. Funct. Mater.* **2008**, *18*, 1047–1056. [CrossRef]
21. Alver, Ü.; Tanrıverdi, A.; Akgül, Ö. Hydrothermal preparation of ZnO electrodes synthesized from different precursors for electrochemical supercapacitors. *Synth. Met.* **2016**, *211*, 30–34. [CrossRef]
22. Zhou, X.; Ma, L. MnO$_2$/ZnO porous film: Electrochemical synthesis and enhanced supercapacitor performances. *Thin Solid Films* **2015**, *597*, 44–49. [CrossRef]
23. Ding, J.J.; Wang, M.Q.; Deng, J.P.; Gao, W.Y.; Yang, Z.; Ran, C.X.; Zhang, X.Y. A comparison study between ZnO nanorods coated with graphene oxide and reduced graphene oxide. *J. Alloy. Compd.* **2014**, *582*, 29–32. [CrossRef]
24. Cai, D.P.; Huang, H.; Wang, D.D.; Liu, B.; Wang, L.L.; Liu, Y.; Li, Q.H.; Wang, T.H. High-Performance Supercapacitor Electrode Based on the Unique ZnO@Co$_3$O$_4$ Core/Shell Heterostructures on Nickel Foam. *ACS Appl. Mat. Interfaces* **2014**, *6*, 15905–15912. [CrossRef] [PubMed]
25. Purushothaman, K.K.; Priya, V.S.; Nagamuthu, S.; Vijayakumar, S.; Muralidharan, G. Synthesising of ZnO nanopetals for supercapacitor applications. *Micro Nano Lett.* **2011**, *6*, 668–670. [CrossRef]
26. Borysiewicz, M.A.; Ekielski, M.; Ogorzalek, Z.; Wzorek, M.; Kaczmarski, J.; Wojciechowski, T. Highly transparent supercapacitors based on ZnO/MnO$_2$ nanostructures. *Nanoscale* **2017**, *9*, 7577–7587. [CrossRef] [PubMed]
27. Hou, S.C.; Zhang, G.H.; Zeng, W.; Zhu, J.; Gong, F.L.; Li, F.; Duan, H.G. Hierarchical core-shell structure of ZnO nanorod@NiO/MoO$_2$ composite nanosheet arrays for high-performance supercapacitors. *ACS Appl. Mater. Interfaces* **2014**, *6*, 13564–13570. [CrossRef]
28. Wang, X.B.; Hu, J.J.; Liu, W.D.; Wang, G.Y.; An, J.; Lian, J.S. Ni-Zn binary system hydroxide, oxide and sulfide materials: synthesis and high supercapacitor performance. *J. Mater. Chem. A* **2015**, *3*, 23333–23344. [CrossRef]
29. Borchert, H.; Haubold, S.; Haase, M.; Weller, H.; McGinley, C.; Riedler, M.; M¨oller, T. Investigation of ZnS passivated InP nanocrystals by XPS. *Nano Lett.* **2002**, *2*, 151–154. [CrossRef]
30. Shi, M.J.; Cui, M.W.; Kang, L.T.; Li, T.T.; Yun, S.; Du, J.; Xua, S.D.; Liu, Y. Porous Ni$_3$(NO$_3$)$_2$(OH)$_4$ nano-sheets for supercapacitors: Facile synthesis and excellent rate performance at high mass loadings. *Appl. Surf. Sci.* **2018**, *427*, 678–686. [CrossRef]
31. Liu, M.C.; Kong, L.B.; Lu, C.; Ma, X.J.; Li, X.M.; Luo, Y.C.; Kang, L. Design and synthesis of CoMoO$_4$-NiMoO$_4$ center dot xH$_2$O bundles with improved electrochemical properties for supercapacitors. *J. Mater. Chem. A* **2013**, *1*, 1380–1387. [CrossRef]

32. Zhang, S.; Pang, Y.; Wang, Y.; Dong, B.; Lu, S.; Li, M.; Ding, S. NiO nanosheets anchored on honeycomb porous carbon derived from wheat husk for symmetric supercapacitor with high performance. *J. Alloy. Compd.* **2018**, *735*, 1722–1729. [CrossRef]
33. Lin, J.; Liu, Y.; Wang, Y.; Jia, H.; Chen, S.; Qi, J.; Qu, C.; Cao, J.; Fei, W.; Feng, J. Designed formation of NiO@C@Cu$_2$O hybrid arrays as battery-like electrode with enhanced electrochemical performances. *Ceram. Int.* **2017**, *43*, 15410–15417. [CrossRef]
34. Li, Q.; Li, C.L.; Li, Y.L.; Zhou, J.J.; Chen, C.; Liu, R.; Han, L. Fabrication of hollow n-doped carbon supported ultrathin NiO nanosheets for high-performance supercapacitor. *Inorg. Chem. Commun.* **2017**, *86*, 140–144. [CrossRef]
35. Xi, S.; Zhu, Y.; Yang, Y.; Jiang, S.; Tang, Z. Facile Synthesis of Free-Standing NiO/MnO$_2$ Core-Shell Nanoflakes on Carbon Cloth for Flexible Supercapacitors. *Nanoscale Res. Lett.* **2017**, *12*, 171. [CrossRef] [PubMed]
36. Yang, P.; Xiao, X.; Li, Y.; Ding, Y.; Qiang, P.; Tan, X.; Mai, W.; Lin, Z.; Wu, W.; Li, T.; et al. Hydrogenated ZnO core-shell nanocables for flexible supercapacitors and self-powered systems. *ACS Nano* **2013**, *7*, 2617–2626. [CrossRef] [PubMed]
37. Yang, H.; Xu, H.; Li, M.; Zhang, L.; Huang, Y.; Hu, X. Assembly of NiO/Ni(OH)$_2$/PEDOT Nanocomposites on Contra Wires for Fiber-Shaped Flexible Asymmetric Supercapacitors. *ACS Appl. Mater. Interfaces* **2016**, *8*, 1774–1779. [CrossRef] [PubMed]
38. Yang, X.J.; Sun, H.M.; Zan, P.; Zhao, L.J.; Lian, J.S. Growth of vertically aligned Co$_3$S$_4$/CoMo$_2$S$_4$ ultrathin nanosheets on reduced graphene oxide as a high-performance supercapacitor electrode. *J. Mater. Chem. A* **2016**, *4*, 18857–18867. [CrossRef]
39. Zhang, G.; Ren, L.; Deng, L.J.; Wang, J.; Kang, L.; Liu, Z.H. Graphene—MnO$_2$ nanocomposite for high-performance asymmetrical electrochemical capacitor. *Mater. Res. Bull.* **2014**, *49*, 577–583. [CrossRef]
40. Ning, P.; Duan, X.; Ju, X.; Lin, X.; Tong, X.; Xi, P.; Wang, T.; Li, Q. Facile synthesis of carbon nanofibers/MnO$_2$ nanosheets as high-performance electrodes for asymmetric supercapacitors. *Electrochim. Acta* **2016**, *210*, 754–761. [CrossRef]
41. Attias, R.; Sharon, D.; Borenstein, A.; Malka, D.; Hana, O.; Luski, S.; Aurbach, D. Asymmetric Supercapacitors Using Chemically Prepared MnO2 as Positive Electrode Materials. *Electrochem. Soc.* **2017**, *164*, 2231–2237. [CrossRef]
42. Wang, Y.; Zhang, D.; Lu, Y.; Wang, W.; Peng, T.; Zhang, Y.; Guo, Y.; Wang, Y.; Huo, K.; Kim, J.; Luo, Y. Cable-like double-carbon layers for fast ion and electron transport: An example of CNT@NCT@MnO$_2$ 3D nanostructure for high-performance supercapacitors. *Carbon* **2019**, *143*, 335–342. [CrossRef]
43. Zhu, F.; Liu, Y.; Yan, M.; Shi, W. Construction of hierarchical FeCo$_2$O$_4$@MnO$_2$ core-shell nanostructures on carbon fibers for high-performance asymmetric supercapacitor. *J. Colloid Interface Sci.* **2018**, *512*, 419–427. [CrossRef]
44. He, Y.; Chen, W.; Li, X.; Zhang, Z.; Fu, J.; Zhao, C.; Xie, A. Freestanding three-dimensional graphene/MnO$_2$ composite networks as ultralight and flexible supercapacitor electrodes. *ACS Nano* **2013**, *7*, 174–182. [CrossRef] [PubMed]
45. Li, L.; Hu, Z.A.; An, N.; Yang, Y.Y.; Li, Z.M.; Wu, H.Y. Facile synthesis of MnO$_2$/CNTs composite for supercapacitor electrodes with long cycle stability. *J. Phys. Chem.* **2014**, *118*, 22865–22872. [CrossRef]

© 2019 by the authors. Licensee MDPI, Basel, Switzerland. This article is an open access article distributed under the terms and conditions of the Creative Commons Attribution (CC BY) license (http://creativecommons.org/licenses/by/4.0/).

Article

Fabrication of GaOx Confinement Structure for InGaN Light Emitter Applications

Yi-Yun Chen [1], Yuan-Chang Jhang [1], Chia-Jung Wu [1], Hsiang Chen [2], Yung-Sen Lin [3] and Chia-Feng Lin [1,*]

[1] Department of Materials Science and Engineering, Innovation and Development Center of Sustainable Agriculture, Research Center for sustainable energy and Nanotechnology, National Chung Hsing University, No. 145, Xingda Road, South Dist., Taichung 402, Taiwan; singyi.sky@gmail.com (Y.-Y.C.); d9883122@gmail.com (Y.-C.J.); thulongwu@gmail.com (C.-J.W.)

[2] Department of Applied Materials and Optoelectronic Engineering, National Chi Nan University, No. 1, University Road, Puli Township, Nantou County 545, Taiwan; hchen@ncnu.edu.tw

[3] Department of Chemical Engineering, Feng Chia University, No. 100, Wenhwa Road, Seatwen, Taichung 40724, Taiwan; yslin@fcu.edu.tw

* Correspondence: cflin@dragon.nchu.edu.tw; Tel.: +886-2284-0500 (ext. 706)

Received: 29 September 2018; Accepted: 4 November 2018; Published: 7 November 2018

Abstract: An indium gallium nitride (InGaN) light-emitting diode (LED) with an embedded porous GaN reflector and a current confined aperture is presented in this study. Eight pairs of n^+-GaN:Si/GaN in stacked structure are transformed into a conductive, porous GaN/GaN reflector through an electrochemical wet-etching process. Porous GaN layers surrounding the mesa region were transformed into insulating GaOx layers in a reflector structure through a lateral photoelectrochemical (PEC) oxidation process. The electroluminescence emission intensity was localized at the central mesa region by forming the insulating GaOx layers in a reflector structure as a current confinement aperture structure. The PEC-LED structure with a porous GaN reflector and a current-confined aperture surrounded by insulating GaOx layers has the potential for nitride-based resonance cavity light source applications.

Keywords: InGaN; porous GaN; insulating GaOx; current confinement aperture structure

1. Introduction

Gallium nitride (GaN) materials have been developed for optoelectronic devices such as light-emitting diodes (LEDs), laser diodes (LD) [1], and vertical cavity surface emitting lasers (VCSEL) [2]. Lateral oxidation of high aluminum (Al) content AlGaAs-based epilayers [3,4] produces a current-confining aperture in the central microcavity structure. The lateral oxidized process was used to transform AlAs into insulating AlOx, acting as a current confinement aperture in the microcavity structure [5]. For nitride materials, Dorsaz et al. [6] reported the use of the electrochemical oxidation process for AlInN to form current apertures. The epitaxial AlGaN/GaN stack [7], AlN/GaN stacks [8,9], and AlInN/GaN stack [10] structures have been reported for bottom epitaxial distributed Bragg reflector (DBR) in GaN-based VCSEL devices. Here, an InGaN-based LED structure with n-GaN:Si/GaN stack epi-layers is transformed into a conductive porous-GaN/GaN reflector through a wet electrochemical (EC) etching process. Then, the porous GaN layers surrounding the mesa region are transformed into insulating GaOx layers in the stack structure through a photoelectrochemical (PEC) oxidation process. A current-confined aperture in the bottom porous GaN reflector was produced in the central mesa region.

2. Experimental

The LED epitaxial layer consisted of a 30 nm-thick GaN buffer layer grown at 530 °C, a 2.0 μm-thick unintentionally doped GaN layer (u-GaN, 1050 °C, 5×10^{16} cm^{-3}), a 1.0 μm-thick n-GaN layer (1050 °C, 2×10^{18} cm^{-3}) for the bottom n-type contact layer, eight pairs of n$^+$-GaN:Si/GaN (n$^+$-GaN:Si with 2×10^{19} cm^{-3}) stack structure, a 0.2 μm-thick u-GaN layer (1050 °C), a 0.3 μm-thick n-GaN layer (1050 °C, 2×10^{18} cm^{-3}), six pairs of In$_{0.2}$GaN/GaN (3 nm/12 nm) multiple quantum wells (MQWs, 760 °C), and a 0.1 μm-thick p-GaN layer (950 °C, 1×10^{18} cm^{-3}). A 150 nm-thick indium tin oxide (ITO) film was deposited on the mesa region and acted as a transparent conductive layer. The mesa regions of the InGaN LED structures were defined using a laser scribing (LS) process and dry etching process with a 1.5 μm etching depth to produce the as-grown eight-period n$^+$-GaN/GaN stack structure. The LED chips were 35×35 μm^2 in size, and the mesa regions were defined by using a triple frequency ultraviolet Nd:YVO$_4$ (355 nm) laser for the front side laser scribing process with a 40 μm spacing width. The samples were immersed in a 0.5 M nitride acid solution for wet electrochemical (EC) etching with an external direct current (DC) bias voltage of +8 V. The EC-treated LED structure with a porous GaN reflector was defined as an EC-LED. Then, the samples were oxidized through a PEC oxidation process with a +20 V bias voltage and illuminated with a 400 W Hg lamp in deionized water for 30 min [11,12]. The porous GaN/GaN stack reflector structure was transformed into a GaOx/GaN stack reflector. The PEC-treated LED with GaOx/GaN stack structure surrounding the mesa region was defined as a PEC-LED. The electroluminescence (EL) spectra were measured through an optical spectrum analyzer (iHR550, Edison, NJ, USA) and the light intensity profiles were measured by a beam profiler (Spiricon, Jerusalem, Israel).

3. Results and Discussions

The optical microscopy (OM) images of the standard-LED (ST-LED), electrochemical etch LED (EC-LED), and photoelectrochemical oxidized LED (PEC-LED) structures are shown in Figure 1a–c, respectively. After the EC wet etching process, the n$^+$-GaN:Si/u-GaN stack structure was transformed into a porous GaN/GaN reflector as shown in Figure 1b. The n$^+$-GaN:Si layers in the stack structure were etched to form the porous GaN layers. The laser scribing line patterns provided the wet etching channels on the n$^+$-GaN:Si/GaN stack structure. The porous GaN reflector in the blue light range was observed on the mesa region. Then, the porous GaN/GaN reflector was transformed into a GaOx/GaN reflector through the PEC oxidation process. The PEC oxidized process occurred from the LS lines and formed the GaOx layers in the stack structure surrounding the mesa region.

Figure 1. OM images of (**a**) ST-LED, (**b**) EC-LED, and (**c**) PEC-LED were observed with laser scribing line patterns. The mesa regions were 35×35 μm.

In Figure 2a, eight pairs of n$^+$-GaN:Si/GaN stack structures were transformed into a porous GaN/GaN stack structure using the EC etching process. The pair thickness of n$^+$-GaN:Si/GaN was about 102 nm. After the PEC oxidation process, the porous GaN layer was oxidized as a GaOx layer, as shown in Figure 2b. The GaN layers in the PEC-LED structure became thinner compared to in the EC-LED structure, indicating that part of the GaN layer was oxidized during the PEC oxidation process. The porous GaN layers in the stack structure provided a channel for the OH$^-$ ions to oxidize the residual single crystalline GaN in the porous structure. The GaOx/GaN stack structure was formed surrounding the mesa region through the lateral PEC oxidation process.

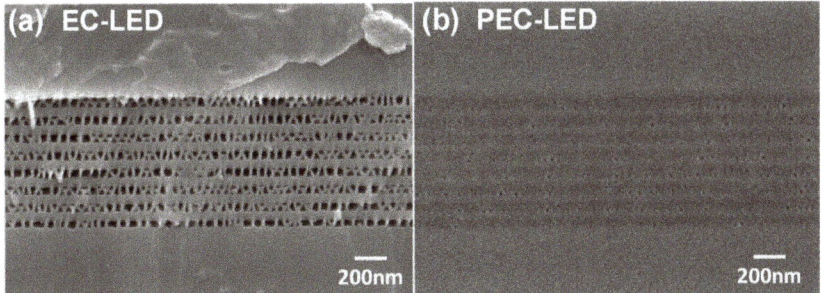

Figure 2. The SEM (scanning electron microscope) micrographs of (**a**) nanoporous GaN/GaN stack structure and (**b**) GaOx/GaN stack structure. The GaOx layer in the PEC-LED structure was prepared through the EC etching and PEC oxidized processes.

The EL (electroluminescence) emission images of the ST-LED, EC-LED, and PEC-LED are provided in Figure 3a–c, respectively, at an 18 µA injection current. In Figure 3a, the uniform emission intensity was observed in the mesa region of the ST-LED. High-EL emission intensity was observed in the EC-LED structure caused by the formation of a high-reflectivity porous GaN reflector (Figure 3b). In the PEC-LED structure, the high-EL emission intensity was localized in the central mesa region as shown in Figure 3c. This occurred because the insulating GaOx layers in the stack structure formed surrounding the mesa region after the lateral PEC oxidation process. The EL light was emitted from the central mesa region and was extracted from the mesa sidewall region. The line EL intensity profiles of the LED samples are observed. After the PEC process, the EL emission intensity of the PEC-LED was localized in the central mesa region. An InGaN-based LED with GaOx current confinement structure was produced via the EC etching process and the PEC oxidation process.

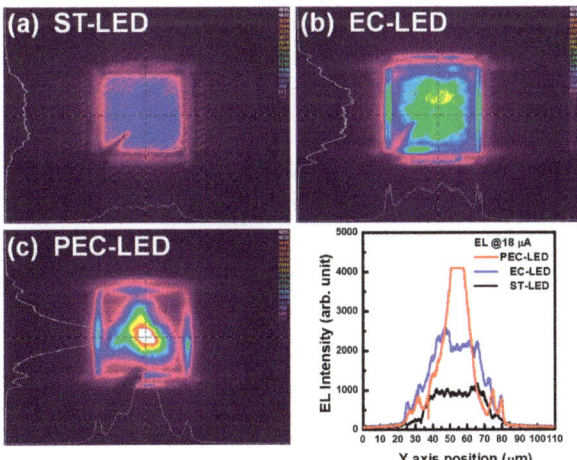

Figure 3. The EL emission images of (**a**) ST-LED, (**b**) EC-LED, and (**c**) PEC-LED observed using a beam profiler. The line intensity profile of the LED samples were measured at an 18 µA operation current (see bottom right Figure).

The reflectance spectra of the LED samples were measured as shown in Figure 4a. The light interference spectrum of the non-treated LED epi-structure was observed due to the light reflection between the top air/GaN:Mg and bottom GaN/sapphire interfaces. The reflectivity of the ST-LED was measured at about 20% as the reference spectrum. In the EC-LED structure, the high reflectance spectrum was observed at 469 nm/80% and 446 nm/76%, which dipped at 456 nm/62%. After the EC etching process, a 0.7 μm-thick InGaN LED cavity was formed between the top air/GaN interface and the bottom porous GaN reflector. The dip wavelength in the reflectance spectrum was caused by the light interference in the short cavity structure. In the PEC-LED structure, the peak reflectance spectrum was observed at 464 nm/69% and 439 nm/66%, which had a dip wavelength at 451 nm/42%. The central dip wavelength shifted from 456 nm (EC-LED) to 451 nm (PEC-LED) due to the transformation process from the porous GaN layers to the GaOx layers in the stack reflector structure. In Figure 4b, the EL peak wavelengths and linewidth were measured at 450 nm and 19.1 nm for the ST-LED, 448 nm/17.2 nm for the EC-LED, and 453 nm/16.5 nm for the PEC-LED structure, respectively, at a 3 mA operation current. The EL emission intensity of the EC-LED was higher than that of the ST-LED, caused by the formation of a highly reflective and conductive porous GaN reflector. The EL peak wavelength of the PEC-LED slightly redshifted compared to ST-LED, potentially caused by the thermal joule heat in the LED device at the high operating current density in the confinement aperture structure.

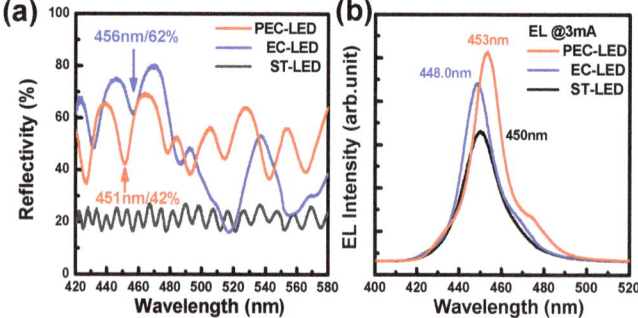

Figure 4. (**a**) The reflectance spectral of all LED structure and (**b**) the EL spectra of the LED samples measured with different embedded reflector spectra.

4. Conclusions

An InGaN LED structure with an embedded porous GaN reflector and a current-confined aperture was fabricated through the lateral EC etching and PEC oxidation processes. The central wavelength of the porous GaN reflector matched the EL emission spectrum. The porous GaN layers provided a large surface area for the following PEC oxidation process. The central wavelength of the reflectors was blueshifted by transforming the porous GaN/GaN stack structure into a GaOx/GaN stack structure. The injection current in the PEC-LED was confined to the aperture structure by being surrounded by insulating GaOx layers in the reflector structure. The produced PEC-LED structure with a porous GaN reflector and a current-confined aperture surrounding by insulating GaOx layers has the potential for nitride-based resonant-cavity LED device applications.

Author Contributions: Y.-Y.C., Y.-C.J. and C.-J.W. fabricated the EC-LED and PEC-LED. H.C. supporting the analysis of the LED samples. Y.-S.L. performed the EL measurement and analysis. C.-F.L. organized the experiment design and wrote the manuscript with contributed from other co-authors.

Funding: This research received no external funding.

Acknowledgments: The authors gratefully acknowledge the financial support for this research by the Ministry of Science and Technology of Taiwan under grant No. 105-2221-E-005-012-MY2 and 107-2221-E-005-061. It was further supported by the Innovation and Development Center of Sustainable Agriculture's Featured Areas Research Center Program within the framework of Taiwan's Ministry of Education's Higher Education Sprout Project.

Conflicts of Interest: The authors declare no conflict of interest.

References

1. Nakamura, S. The Roles of Structural Imperfections in InGaN-Based Blue Light-Emitting Diodes and Laser Diodes. *Science* **1998**, *281*, 956–961. [CrossRef]
2. Someya, T.; Werner, R.; Forchel, A.; Catalano, M.; Cingolani, R.; Arakawa, Y. Room Temperature Lasing at Blue Wavelengths in Gallium Nitride Microcavities. *Science* **1999**, *285*, 1905–1906. [CrossRef] [PubMed]
3. Yang, G.M.; MacDougal, M.H.; Dapkus, P.D. Ultralow Threshold Current Vertical-cavity Surface-emitting Lasers Obtained with Selective Oxidation. *Electron. Lett.* **1995**, *31*, 886–888. [CrossRef]
4. Bissessur, H.K.; Koyama, F.; Iga, K. Modeling of Oxide-confined Vertical-cavity Surface-Emitting Lasers. *IEEE J. Sel. Top. Quantum Electron.* **1997**, *3*, 344–352. [CrossRef]
5. MacDougal, M.H.; Dapkus, P.D.; Bond, A.E.; Lin, C.-K.; Geske, J. Design and Fabrication VCSEL's with Al_xO_y–GaAs DBR's. *IEEE J. Sel. Top. Quantum Electron.* **1997**, *3*, 905–915. [CrossRef]
6. Dorsaz, J.; Bühlmann, H.J.; Carlin, J.F.; Grandjean, N.; Ilegems, M. Selective Oxidation of AlInN Layers for Current Confinement in III–nitride Devices. *Appl. Phys. Lett.* **2005**, *87*, 072102. [CrossRef]
7. Nakada, N.; Nakaji, M.; Ishikawa, H.; Egawa, T.; Umeno, M.; Jimbo, T. Improved characteristics of InGaN multiple-quantum-well light-emitting diode by GaN/AlGaN distributed Bragg reflector grown on sapphire. *Appl. Phys. Lett.* **2000**, *76*, 1804–1806. [CrossRef]
8. Lin, C.F.; Yao, H.H.; Lu, J.W.; Hsieh, Y.L.; Kuo, H.C.; Wang, S.C. Characteristics of Stable Emission GaN-Based Resonant-Cavity Light-Emitting Diodes. *J. Cryst. Growth* **2004**, *261*, 359–363. [CrossRef]
9. Yao, H.H.; Lin, C.F.; Kuo, H.C.; Wang, S.C. MOCVD Growth of AlN/GaN DBR Structures under Various Ambient Conditions. *J. Cryst. Growth* **2004**, *262*, 151–156. [CrossRef]
10. Carlin, J.F.; Ilegems, M. High-Quality AlInN for High Index Contrast Bragg Mirrors Lattice Matched to GaN. *Appl. Phys. Lett.* **2003**, *83*, 668–670. [CrossRef]
11. Lin, C.F.; Yang, Z.J.; Zheng, J.H.; Dai, J.J. Enhanced Light Output in Nitride-based Light-emitting Diodes by Roughening the Mesa Sidewall. *IEEE Photonics Technol. Lett.* **2005**, *17*, 2038–2040.
12. Lin, C.F.; Chen, K.T.; Huang, K.P. Blue Light-Emitting Diodes with an Embedded Native Gallium Oxide Pattern Structure. *IEEE Electron Device Lett.* **2010**, *31*, 1431–1433. [CrossRef]

© 2018 by the authors. Licensee MDPI, Basel, Switzerland. This article is an open access article distributed under the terms and conditions of the Creative Commons Attribution (CC BY) license (http://creativecommons.org/licenses/by/4.0/).

Article

Highly Visible Photoluminescence from Ta-Doped Structures of ZnO Films Grown by HFCVD

Víctor Herrera [1], Tomás Díaz-Becerril [1,*], Eric Reyes-Cervantes [2], Godofredo García-Salgado [1], Reina Galeazzi [1], Crisóforo Morales [1], Enrique Rosendo [1], Antonio Coyopol [1], Román Romano [1] and Fabiola G. Nieto-Caballero [1]

[1] Centro de Investigación en Dispositivos Semiconductores, Universidad Autónoma de Puebla, 14 sur y Av. San Claudio, C.U., Edif. IC-5, Puebla 72570, Mexico; pds_vherrera@hotmail.com (V.H.); godgarcia@yahoo.com (G.G.-S.); ingquim25@gmail.com (R.G.); crisomr@yahoo.com.mx (C.M.); enrique171204@gmail.com (E.R.); acoyopol@gmail.com (A.C.); roman.romano@gmail.com (R.R.); ruscaballero@yahoo.com.mx (F.G.N.-C.)

[2] Centro Universitario de Vinculación y Transferencia de Tecnología, Prol. de la 24 sur esq. con Av. San Claudio, C.U., Edif. CUVyTT, Puebla 72570, Mexico; eric.cervantes@correo.buap.mx

* Correspondence: tomas.diaz.be@gmail.com; Tel.: +52-222-229-5500 (ext. 7871)

Received: 27 July 2018; Accepted: 3 September 2018; Published: 22 October 2018

Abstract: Tantalum-doped ZnO structures (ZnO:Ta) were synthesized, and some of their characteristics were studied. ZnO material was deposited on silicon substrates by using a hot filament chemical vapor deposition (HFCVD) reactor. The raw materials were a pellet made of a mixture of ZnO and Ta_2O_5 powders, and molecular hydrogen was used as a reactant gas. The percentage of tantalum varied from 0 to 500 mg by varying the percentages of tantalum oxide in the mixture of the pellet source, by holding a fixed amount of 500 mg of ZnO in all experiments. X-ray diffractograms confirmed the presence of zinc oxide in the wurtzite phase, and metallic zinc with a hexagonal structure, and no other phase was detected. Displacements to lower angles of reflection peaks, compared with those from samples without doping, were interpreted as the inclusion of the Ta atoms in the matrix of the ZnO. This fact was confirmed by energy dispersive X-ray spectrometry (EDS), and X-ray diffraction (XRD) measurements. From scanning electron microscopy (SEM) images from undoped samples, mostly micro-sized semi-spherical structures were seen, while doped samples displayed a trend to grow as nanocrystalline rods. The presence of tantalum during the synthesis affected the growth direction. Green photoluminescence was observed by the naked eye when Ta-doped samples were illuminated by ultraviolet radiation and confirmed by photoluminescence (PL) spectra. The PL intensity on the Ta-doped ZnO increased from those undoped samples up to eight times.

Keywords: zinc oxide; tantalum oxide; ZnO:Ta doped films; substitutional alloy

1. Introduction

In recent years, ZnO has become one of the most closely studied materials, due to its very interesting properties in optoelectronic devices applications, such as room temperature lasers [1], light emitting diodes [2,3], ultraviolet (UV) detectors [4], field-emission displays [5,6], photonic crystals [7], solar cells [8–10], and sensing in the nano-size range [11,12]. The control of its properties is critical in the context of novel applications. For this reason, a better understanding of the synthesis of microstructures/nanostructures with different morphologies is important and necessary to achieve objectives and applications on materials for the modern world [13].

Several techniques have been employed to make ZnO in thin films and nano- or microstructures; among them are radio-frequency magnetron sputtering, pulsed laser deposition, spray pyrolysis, and thermal evaporation flame synthesis [14–17]. The properties of the materials seem to be dependent on the process and technique used to synthesize them. Developing new or modified methods to obtain those type of materials could report new insights on topics that are not yet understand for those types of oxides.

A profitable, simple, and easily scalable method to synthesize the material the hot filament chemical vapor deposition (HFCVD). This technique allows us to produce multifunctional structures on a larger scale. In addition, it is possible to control the material properties by only changing the synthesis conditions.

This technique has been used to grow a wide variety of films, including diamond [18], as well as amorphous silicon nitride [19,20], semiconductor compounds [21], and more. Recently, the technique has been used to grow several nanostructures of different materials, such as silicon-rich oxides [22], nanocrystalline silicon [23], graphene [24], molybdenum selenide [25], silicon carbide [26], and zinc sulfide [27] among others, for applications in new or improved devices, such as solar cells, sensors, and metal oxide semiconductors (MOS) structures [18,23]. HFCVD technology is compatible with the current silicon-based technology, and it has advantages over other methods because it can produce materials at low cost, and on a wide substrate surface.

An important issue regarding applications in electronic and optoelectronic technologies is doping. ZnO exhibit naturally n-type conductivity because the presence of punctual defects, such as zinc interstitial and/or oxygen vacancies in its matrix [28,29]. P-type is not easy to obtain, due to the compensation effect that impedes this [3,29]. However, some other properties, such as optical emission can be changed when ZnO is doped. It is reported light emission, in the UV-visible range, can be obtained when the material is doped with transition metals.

It has been reported that green emission occurs at around 510 nm when copper atoms are incorporated in ZnO films [30–32]. The luminescence was ascribed mainly to the existence of oxygen vacancies. Other transition metals when introduced to ZnO, such as Al, Li, Co, Mn, and Zn, have also been reported to exhibit emissions in different ranges; however, the wavelengths of emission and intensity depended on the process and technique of the growth [10,33–35]. Similarly, yellow emission was been observed when the ZnO synthesis was performed by chemical techniques. This emission was assigned to the existence of oxygen interstitials [36].

Ta is an atom that has been introduced in ZnO to change the properties of the material. It was found that the Ta atoms can substitute Zn atoms in the ZnO matrix, modifying the electrical and optical properties of the films. Some uses include photoactinic and TCO (transparent and conductive oxide) uses, among others. However, information on the changes on the photoluminescent properties with the incorporation of Ta in the ZnO material is missing.

To investigate on the effect of ZnO doping with Ta atoms, core-shell structures doped with tantalum were fabricated by the HFCVD method, and the optical (photoluminescence), structural, morphological, and electrical properties were studied and reported. By comparison, the undoped structures were also prepared, to observe the effect of the dopant.

The results showed that the properties of ZnO core-shell structures are a bit different from those of pure ZnO crystals. Photoluminescence (PL) intensity on Ta-doped ZnO increased from those undoped samples by up to eight times. This effect was ascribed to a change in the morphology of the structures, with the incorporation of tantalum in the process. The morphology of the structures in this work was found to be modified with the incorporation of tantalum in the process; this effect increases the specific area of the material, increasing the number of radiative centers.

Tantalum atoms do not appear to introduce additional radiative centers that contribute to the PL, but they affect the electricals of the ZnO grains. Ta is a donor in ZnO that contributes to reduce the resistivity of the ZnO grains in the shell. The results are in concordance with other authors.

2. Materials and Methods

Films obtained in this work were grown using a hot filament chemical vapor deposition system (HFCVD). The experimental homemade reactor consisted of elements showed in Figure 1. As can be seen in the figure, the system has a 2-way valve to introduce a reactant gas into the chamber, such as hydrogen or nitrogen; in this work, molecular hydrogen was only used as reactant to grow the films. Also, the system has a flowmeter and a pressure valve to control gases into the chamber; in this work, the reaction chamber was saturated with hydrogen, and the temperature in the substrate was kept at 800 °C.

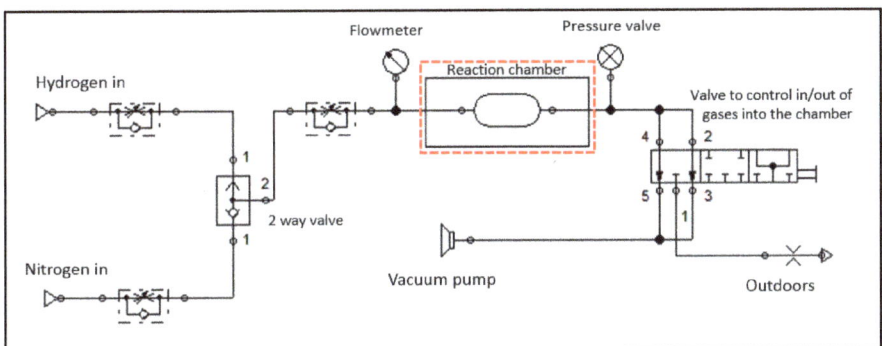

Figure 1. Schematic diagram of the hot filament chemical vapor deposition (HFCVD) homemade system used to grown films in this work.

The reaction chamber (red box in Figure 1) consists of a cylindrical tube of quartz that was selected to support elevated temperatures in its interior. The tube has metal caps of stainless steel and O-rings at each end to ensure complete isolation between the inside of the tube under atmosphere, owing to elevated temperatures, reactant gases are highly pyrophoric. Principal components inside the reaction chamber are shown in Figure 2, and a sprinkler of stainless steel was used to fill the reaction chamber with the reactant gas. A filament of tungsten was used to raise the temperature inside the reaction chamber; in this work, the filament was fixed at 2000 °C. A pellet fabricated with a mixture of powders of ZnO (Mallinckrodt Chemicals CAS 1314-13-2, St. Louis, MO, USA) and Ta_2O_5 (ALDRICH CAS 1314-61-0, St. Louis, MO, USA) was used as a precursor to obtain the ZnO and ZnO:Ta doped films. The amount of ZnO powder was kept in 500 mg, and the amount of tantalum oxide was varied by a percentage of 0%, 10%, 20%, 40%, 50%, and 60%, for 500 mg of ZnO, i.e., 500 mg ZnO + 0 mg Ta_2O_5, 500 mg ZnO + 50 mg Ta_2O_5, 500 mg ZnO + 100 mg Ta_2O_5, and so on. After having the required weights of each powder, both powders of ZnO and Ta_2O_5 were mixed until a homogenous powder was obtained. After that, each mix was 1 ton-pressed. A disk shape of 10 mm in diameter and 5 mm in height were the result. All films were grown on a silicon n-type substrate.

The grown process was as follows: the reactor was sealed, and a flow of hydrogen was passed twice through it; next, the hydrogen flow was maintained inside the chamber and the filament is turned on; this process was kept by 3 min.

At a filament temperature of 2000 °C, the molecular hydrogen gas was dissociated into radicals that were highly reactive, which impinged on the pellet that was used as the source to produce intermediaries, such as Zn (vapor), Ta (vapor), and OH principally, although the additional reactions could occur with the gas and the substrate [37]. Finally, by the reactor dynamics, the gases are transported to the substrate where the film is obtained, optical, and structural properties of this were studied by X-ray, SEM, energy dispersive X-ray spectrometry (EDS); the Hall Effect, and PL techniques.

The crystalline structure and the orientation of the crystallites were studied using X-ray diffraction (BRUKER D8 with Cu Kα radiation (λ = 1.541 Å), Billerica, MA, USA); Raman spectra were taken with a RAM HR800 Raman spectrometer (HORIBA, Les Ulis, France) equipped with an Olympus BX41 microscope (Olympus, Tokyo, Japan); a 632.8 nm He-Ne laser was used as the excitation source. The surface morphology and composition of the films were examined with two systems; the low resolution was taken with a system SEM Jeol model 6610LV (JEOL Ltd., Tokyo, Japan) equipped with an INCA energy dispersive X-ray attached to this, and the high resolution was analyzed with a field emission scanning electron microscope, FESEM JSM 5400LV (JEOL Ltd., Tokyo, Japan) equipped with a NORAN energy dispersive X-ray spectrometer. The profilometry and roughness analyses of the films were taken with a surface profiler Dektak 150 (Veeco Instruments, Inc., Plainview, NY, USA). Photoluminescence measurements were performed at room temperature with a spectrofluorometer FluroMax 3 (Horiba, Ltd., Kyoto, Japan) that was equipped with an emission detector with high sensibility and a 150 W xenon lamp to excite the samples. The electrical properties of the films were determined with a Hall Effect system from Ecopia Company, model HMS-5000 (Ecopia, Waterloo, ON, Canada), using a magnetic field of 0.5 T.

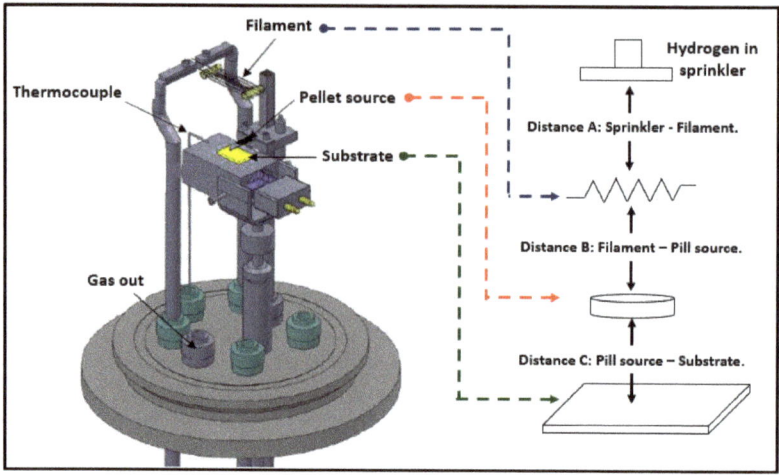

Figure 2. The principal components inside of the reaction chamber.

3. Results and Discussion

3.1. Structural Characterization by XRD

The diffractograms of the films of polycrystalline zinc oxide doped with tantalum are shown in Figure 3, and a pure spectrum of ZnO is presented as a reference. For the pure ZnO material (Figure 3a), the presence of the diffraction peaks indicated that the material possessed a polycrystalline structure. Three well-distinguished diffraction peaks placed at $2\theta = 32°$, $34.6°$, and $36.4°$, corresponding to planes (100), (002), and (101) were identified and matched with a hexagonal (wurtzite) ZnO crystalline structure (pdf card 075-0576). Additionally, some other diffractions peaks could be seen at $2\theta = 47.8°$, $56.8°$, $63.1°$, $68.0°$, and $69.3°$, corresponding to reflections on (102), (110), (103), (112), and (201) planes of wurtzite ZnO structure respectively. The presence of the Zn phase was identified by the peaks placed at $2\theta = 39.2°$, $43.3°$, and $54.5°$, corresponding to planes (100), (101), and (102) of Zn in the hexagonal phase (pdf card 004-0831).

The diffraction peaks from the ZnO:Ta samples are presented in Figure 3b–f. Some differences can be seen from that of reference sample (Figure 3a). As can be observed, the peak of ZnO in $2\theta = 69.3°$ was could not peak in the other range; to get a better view of other peaks, we plotted the same diffraction patterns in a range from $2\theta = 30°$ to $66°$ in Figure 4. In this graphs, it was possible to observe an additional diffraction peak arising at $2\theta = 66.98°$ corresponding to the structure of $ZnTa_2O_6$, suggesting some that Ta atoms were incorporated into the ZnO lattice (pdf card 049-0746) [38]. Because the atomic radii of Ta (0.78 Å) are very similar to Zn atoms (0.82 Å) these atoms easily replace some zinc atoms in the ZnO network without making abrupt changes in the ZnO lattice. In us case, a small level expansive stress in the ZnO lattice can be observed in $2\theta = 63.1$, corresponding to the plane (103); this stress caused this peak to shift to lower angles (red arrows in Figure 4c,d), but this peak could not be observed clearly in all diffractions [39,40].

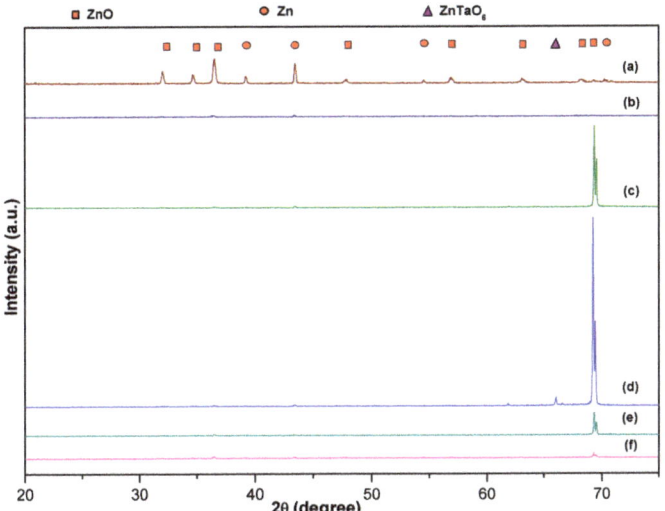

Figure 3. X-ray diffraction (XRD) patterns for (**a**) undoped ZnO film, and ZnO:Ta-doped films grown with a pellet weight of (**b**) 50 mg, (**c**) 100 mg, (**d**) 200 mg, (**e**) 250 mg, and (**f**) 300 mg of Ta_2O_5.

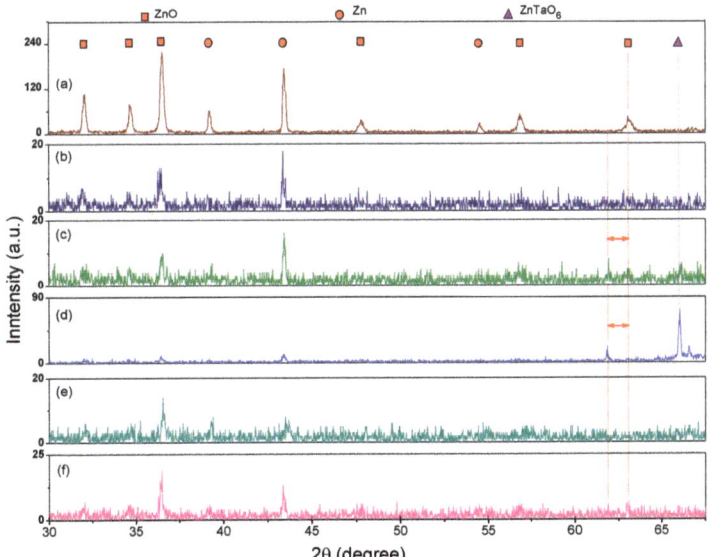

Figure 4. Patterns for (**a**) undoped ZnO film, and ZnO:Ta-doped films grown with a pellet weight of (**b**) 50 mg, (**c**) 100 mg, (**d**) 200 mg, (**e**) 250 mg, and (**f**) 300 mg of Ta_2O_5, in a range of $2\theta = 30°$ to $66°$.

It is interesting to note that for the principal peak of ZnO at $2\theta = 36.4°$ in Ta-doped samples corresponding to (101), the plane changed strongly in intensity as the concentration of Ta was increased in the process (Figure 4). This result suggests that the high content of Ta may lead to the degradation of ZnO crystallinity. The other ZnO peaks also became small and broad, which could be an indication that tantalum atoms in the HFCVD process could modify the crystalline orientation of the growth, and from here modify the film morphology.

Diffraction peaks were analyzed with software HighScore Plus of PANalytical, every profile was approximated using a standard Caglioti function with a Pseudo-Voight profile fitting; the results are shown in Table 1.

Table 1. Lattice parameters and crystallite size for ZnO and ZnO:Ta-doped films obtained.

Reference Name	d (Å)	(hkl)	a (Å)	c (Å)	Average Crystallite Size (nm)
ZnO	2.5869	(002)	3.2406	5.1738	224
ZnO 50 mg Ta	2.4653	(101)	3.2361	5.1841	155
ZnO 100 mg Ta	1.3542	(201)	3.2401	5.1806	142
ZnO 200 mg Ta	1.3561	(201)	3.2448	5.1831	134
ZnO 250 mg Ta	1.3546	(201)	3.2407	5.1874	129
ZnO 300 mg Ta	1.3551	(201)	3.2419	5.1895	122

Analyzing the diffraction chart, it can be see the three important diffraction peaks of (100), (002), and (101), the planes of ZnO were diminished, with the introduction of Ta in the ZnO process, which it could mean the growth in that direction is inhibited while the growth along (201) is favored. According to the data obtained and presented in Table 1, it can be seen that the crystallite size of ZnO is bigger than the crystallites of ZnO:Ta doped, because when ZnO is doped with tantalum atoms, the crystallite size decrease strongly, and after this, the crystallite size decreases slowly, in this case, this fact could be an indication that the tantalum atoms are creating this effect.

According to the other reviewed reports, when ZnO is doped with tantalum, the crystallite size is modified, but there is no consensus of the increment or decrement in the crystallite size in ZnO:Ta-doped films; this fact seems to be more greatly affected by the technique used to grow the film; even when ZnO is doped with other materials, the crystallite size sometimes increases, while at other times it decreases [38,40,41]. For example, Bang et al. reported the effect of lithium doping in ZnO films, they found that the (002) planes of ZnO were reduced as the Li content was increased; this was interpreted as a degradation of the ZnO crystallinity, while an increase in the grain size was observed as the concentration of lithium increased [42]. Also, it has been reported that the presence of metallic zinc is associated with the existence of the core-shell structures, where the metallic zinc is the core and the ZnO nanocrystals are comprise the shell [37].

3.2. Raman Measurements

The room temperature Raman spectra for undoped and Ta-doped ZnO in the range of 100 to 1000 cm^{-1} are presented in Figure 5. Three main bands were observed for pure ZnO (Figure 5a). A peak labeled E_2 (High) at 437 cm^{-1} is known as the active optical phonon mode. The vibrational modes at 97 cm^{-1}, known as E_2 (LO) and 570 cm^{-1}, known as A_1 (LO), are also observable on the spectra. The presence of the three observed vibrations, in our measurements, confirmed that the material synthesized had wurtzite hexagonal structure [43–46].

It was reported that a broad Raman band at about 570 cm^{-1} may originate from defects related to the phonon mode, O-vacancy-related stable complexes [47], zinc interstitials, or from the disorder-activated B1 (High) silent mode [48] of w-ZnO.

It is well known that when metallic zinc particles are exposed to air or/and to an oxygen environment, a $Zn_{1+X}O$-defective zinc oxide with a Zn ion in the interstitial position is formed at around the metallic zinc [49], and a broad Raman band with a peak centered at 560 cm^{-1} was seen on the spectra. If it was assumed that the Raman band centered at 563 cm^{-1}, as seen in Figure 5a, had a contribution due to the formation of a defective oxide, then this result suggests us that our material was a core-shell type, where the core was metallic zinc and the shell consisted of ZnO grains.

Moreover, the undoped and Ta-doped ZnO samples were grown under an oxygen-deficient environment, and therefore, the synthesized material contained a great number of oxygen vacancies and zinc interstitials [29,50,51], and they could also contribute to the broadening of the Raman band observed at 563 cm^{-1}.

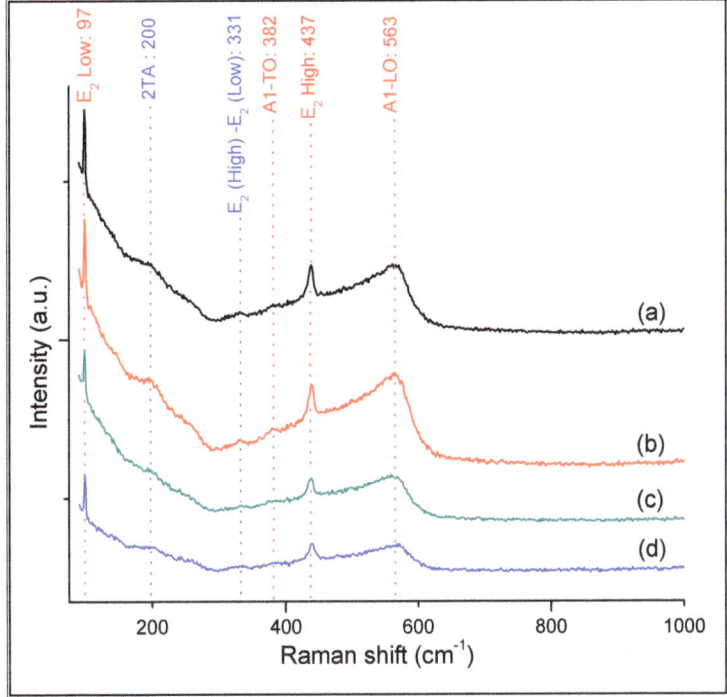

Figure 5. Raman shift from the samples of (**a**) undoped ZnO, and ZnO:Ta-doped films, grown with pellet weight of (**b**) 50 mg, (**c**) 200 mg, and (**d**) 300 mg of Ta_2O_5.

Raman spectra from the Ta-doped ZnO are shown in Figure 5b–d. It can be seen that all of the peaks were found to correspond to that from the undoped sample (Figure 5a); the only difference was the reduction in the intensity of the E_2 (high) vibration, and the broadening of both the second order and A_1 (LO) modes.

This observation could be an indication that Ta is encrusted in the ZnO matrix, and that it induces small disorder in the lattice. The reduced intensity in the E_2 (high) peak might be related to the degradation on the crystalline quality as observed by X-ray diffraction; we can expect those tantalum atoms, when incorporated in the film, to induce change in the bonds strength between zinc and oxygen; therefore crystallinity is affected by this process, as have been observed in XRD results.

3.3. SEM and EDS Studies

Analysis by SEM was carried out to observe the film morphology, and the images are presented in Figure 6. It can be seen the surface morphology changed as the percentage of Ta_2O_5 increased in the pellet source. The picture for undoped ZnO film (Figure 6a) shows a morphology with a high density of big spherical shapes, and both the density and size of these structures reduced or even disappeared as the Ta_2O_5 content increased in the process (Figure 6b–f). Magnification of these samples (Figure 7) showed the shape details of their morphology.

Micrographs taken on the Ta-doped ZnO samples (Figure 6b–f) showed that the roughness of the surface decreased as the content of Ta_2O_5 was increased in the pellet source; measurements of profilometry were taken over the surface of each film, and the results obtained showed an irregular surface, as can be seen in that Figure 6; the results obtained showed depths with a minimum of 10 μm and a maximum of up to 150 μm, but surface roughness (RMS) analysis performed with this technique showed that surface tended to form a softer surface when the amount of tantalum was greater in the growth process; this RMS factor was plotted and it can be observed in Figure 8.

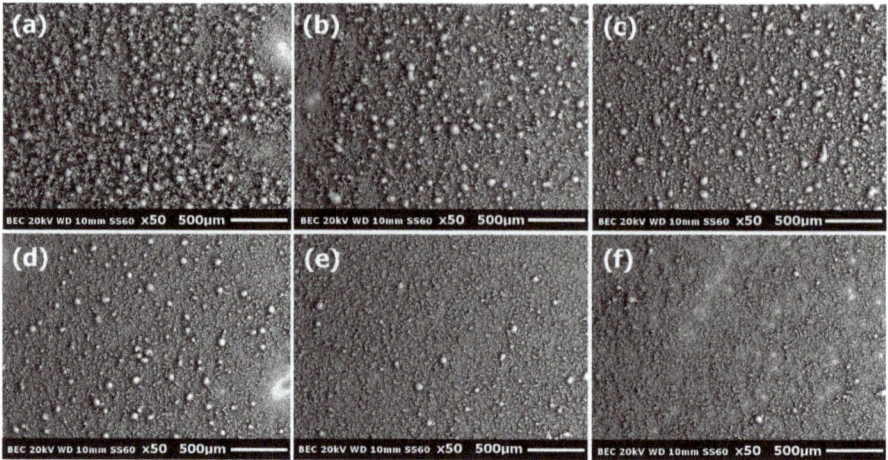

Figure 6. Results of scanning electron microscopy (SEM) 50× of (**a**) undoped ZnO film, and ZnO:Ta-doped films grown with a pellet weight of (**b**) 50 mg, (**c**) 100 mg, (**d**) 200 mg, (**e**) 250 mg, and (**f**) 300 mg of Ta_2O_5.

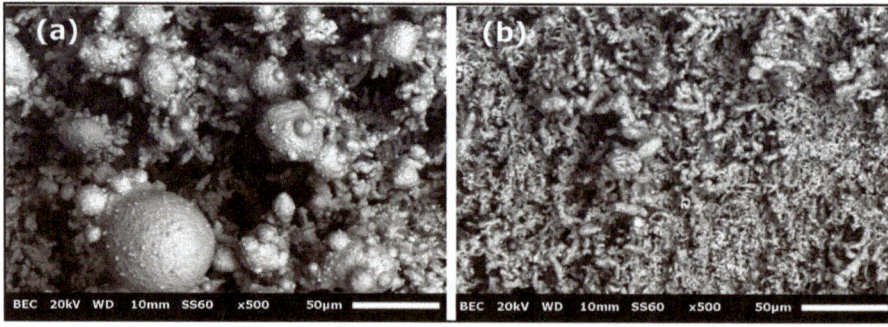

Figure 7. Zoom of 500× on the film of (**a**) ZnO, and (**b**) ZnO:Ta doped with 300 mg of Ta_2O_5 on pellet.

A detailed SEM image from those samples with a 500× zoom is shown in Figure 7. The surface was composed by spheres and rods in a random form like a porous sponge. Additionally, a FESEM analysis of Figure 7a and b displayed the same type of spherical structures, but of different size. Some of those structures seemed hollow or even broken, showing a shell type structure, this shapes could be observed with a high resolution for ZnO in Figure 9(a.1–a.3), and for ZnO:Ta doped in Figure 9(b.1–b.3).

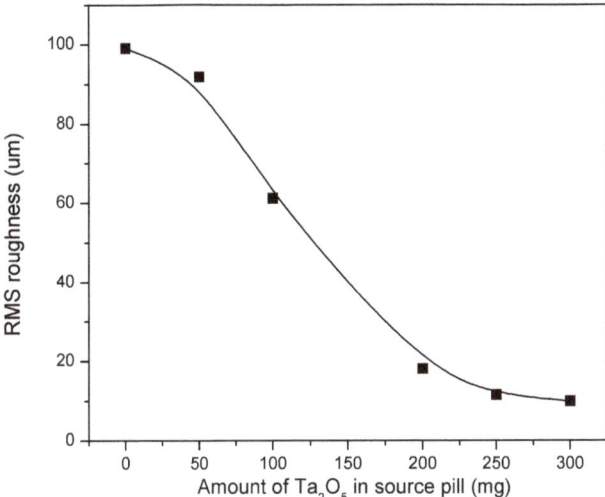

Figure 8. Roughness dependence in films obtained according to the amount of Ta in the source pellet.

Figure 9. High-resolution field emission (FE) SEM measurements for (**a.1–a.3**) undoped ZnO film, and (**b.1–b.3**) ZnO:Ta-doped film in different zones of scanning, grown with 300 mg of Ta_2O_5 in the pellet.

From EDS analysis carried out on undoped ZnO films, only oxygen, and zinc elements were detected (Figure 10a). The relation of Zn/O of greater than one indicated that the material was out of stoichiometry, which meant a zinc-rich ZnO material. Our results were in agreement with those found by other authors for this type of systems [37,52,53].

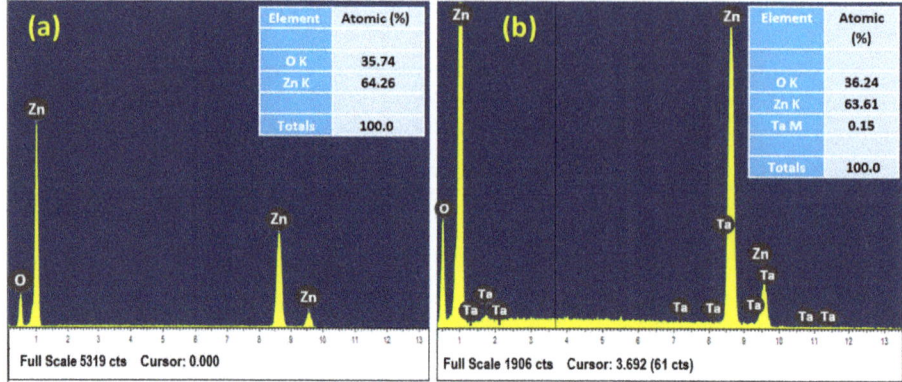

Figure 10. X-ray spectrometry (EDS) analysis for (**a**) undoped ZnO film, and (**b**) ZnO:Ta-doped film grown with 300 mg of Ta_2O_5 in the pellet source.

EDS results on Ta-doped ZnO samples (Figure 10b) indicate that it was a Zn-rich ZnO material, but there was also a clear contribution of the tantalum element around of 0.1% and 0.15% in atomic percentage (±0.5% in weight), this could be owing to the fact that tantalum atoms are mainly incorporated inside the lattice of ZnO, but no over surface of the structures obtained.

The data above could shed some light about the growth process. For undoped ZnO samples, the spheres could be formed as follows. The atomic hydrogen impacted on the pellet source producing OH^- (gas) and Zn (gas) according to the reaction proposed in [37,39]. After that, Zn atoms condensed in a vapor phase, and some Zn liquid drops were formed and they acted as nucleation sites, and OH ions are attached to the surface, forming a mixture of zinc in the liquid state, covered with a thin film of ZnO (solid), and they were deposited on the substrate, where it eventually coalesced and this drop increased in size. Further, due to the presence of OH, the zinc oxide cover increases in thickness, and a core(liquid)–shell(solid) structure was formed because of differences in the melting points 420 °C and 1975 °C for zinc and ZnO respectively.

As the growth was performed at a high temperature, the nuclei remained in a liquid state, producing a Zn gas pressure, due to the vapor–liquid equilibrium, and the Zn atoms diffused through the grain boundaries of the shell to release the internal pressure. Eventually the shell broke and Zn gas escaped, producing holes on the shell, as seen in Figure 9. This effect has been reported in the literature, and concentric core-shell (Zn/ZnO) balls have been synthesized by oxidizing ZnO powders at low temperatures in air, followed by annealing at temperatures above the zinc melting point [39,54,55].

The existence of this crystalline zinc forming the core were detected by X-ray diffraction results (Figure 3) and by EDS, where it obtained a relation Zn/O of 69% at 31%, this means a zinc-rich ZnO material.

On the other hand, for Ta-doped ZnO samples; the formation of these structures could be explained as a similar form as undoped but instead of forming Zn liquid drops, it forms Zn–Ta alloy liquid drops. Because the difference of the fusion point between Zn (420 °C) and Ta (2996 °C) the properties of the alloy changes and it can affect the size and form of the deposited material, as seen in the sequence in Figure 6b–f. The higher the percentage of Ta_2O_5, the small structures are obtained. The shell is composed of small polyhedral crystals of ZnO material according to SEM, X-ray, and Raman measurements.

The crystals forming the shell were also affected by the introduction of Ta, as can be observed from FESEM in Figure 11b. The ZnO composing the shell had sheet-like polycrystalline shapes with a smaller crystallite size than the undoped ZnO (see Table 1).

The effect of both the decreased grain size and the change in the shape of the crystals in the ZnO:Ta-doped films was also observed and reported by different authors [40,41,56–58], and it was in agreement with our experiments. This effect could be owing to the fact that tantalum atoms create a compression stress inside the lattice of ZnO, and they prevents an increase in these big shapes, as generated by the difference of ionic radii between Zn and Ta, as commented before in the XRD results. In fact, this phenomenon also has been observed when ZnO is doped with tantalum by other techniques [56], or when ZnO is doped with other materials such as aluminum and lithium [35,42].

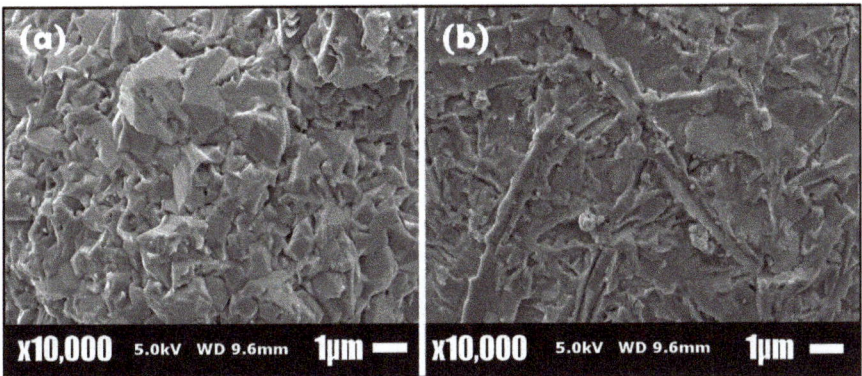

Figure 11. FESEM measurements on the shell from (**a**) undoped ZnO film, and (**b**) ZnO:Ta-doped film obtained with 300 mg of Ta_2O_5 in the pellet.

The synthesis of smaller structures means that somehow, the relation between the surface/volume of the film will increase, and this fact will be reflected in an increment of reactions between the existing species on the surface of structures, and also a possible increment in the adsorption and resorption of the species in these kind of porous films, and moreover, an increment of cations or anions on the surface of the film, as has been reported in other works; in fact, the role of the surface area has been observed to affect various mechanisms such as gas sensors, biological cells, and the effects of fluorescence emission [28,59–61].

3.4. Photoluminescence

The photoluminescence (PL) technique is a useful way to obtain information about the energy states of the impurities and defects of a material. Figure 12 shows the PL properties from our samples. As can be observed in that figure, all films exhibited two main bands, one sharp curve of around 380 nm, and another one with a more widened curve centered around 495 nm and extending from 425 nm to 625 nm. The PL intensity first increased as the percentage of Ta increased, and after that it decreased.

Photoluminescent emissions in ZnO were grouped into two main bands denominating the emissions near-band-edge (NBE), and deep-level emissions (DLE) corresponding to the transition band-to-band and the transition through to the deep centers, with energy levels in the forbidden gap, respectively. Emissions NBE in ZnO have been attributed to free excitation recombination, and they are observed at around the ultraviolet range.

Nevertheless, emissions DLE in ZnO have been assigned to the presence of various point defects, either intrinsic or extrinsic, which introduce deep levels in the bandgap and are assumed to be responsible for PL band broadening in the range between the blue and red emission [3,28,29,34,62].

As can be seen in Figure 12, the main PL band is the DLE band, and it increases in intensity by up to 60% as the Ta_2O_5 is incorporated into the growth process, and after that it reduces. Moreover, all bands are peaked at the same wavelength, independently of the percentage of Ta_2O_5, which means the doping with Ta does not seem to play a role in the generation of radiative deep centers other than for those of pure ZnO.

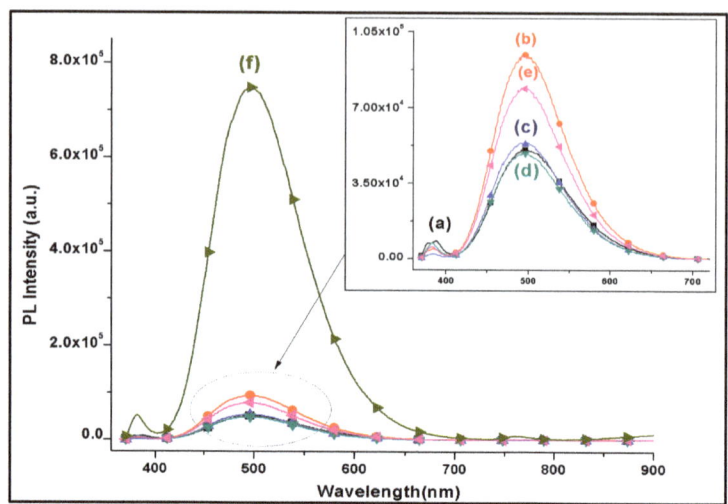

Figure 12. Photoluminescence spectra of (**a**) ZnO film and ZnO:Ta-doped films with (**b**) 10% Ta_2O_5, (**c**) 20% Ta_2O_5, (**d**) 40% Ta_2O_5, (**e**) 50% Ta_2O_5, and (**f**) 60% Ta_2O_5 in the pellet source.

To investigate the possible origin of the green luminescence observed in this type of films, many reports that try to describe this phenomena exactly, were reviewed, but there is no consensus, as the green emission has been observed on several films doped with Cu or Co atoms, on films with zinc vacancies (V_{Zn}), oxygen vacancies (V_O), interstitial zinc ions (Zn_i), oxygen antisites (O_{Zn}), and transitions $Zn_i \rightarrow V_{Zn}$, for these reasons, researchers came to the conclusion that this emission is a combination of several deep levels [62]. For example, in reference [63], it was reported that when ZnO films were annealed at temperatures higher than 800 °C, oxygen vacancies and zinc vacancies are generated, producing green emissions at 490 and 530 nm. In addition, a good correlation between the green emission and singly ionized oxygen vacancies was observed in commercial ZnO powders, annealed in forming gas (N_2:H_2) or O_2 at temperatures ranging from 500 to 1050 °C [64].

By other side, has been reported that either zinc vacancy and oxygen vacancy could contribute to green emission through shallow donors, and even Rodnyi and Khodyuk [62] reported that it is possible to assume the existence of donors with two levels (ground level and excited one) instead of two kinds of shallow donors. It has also been observed that in films of ZnO with an excess of oxygen, the maximum of the green luminescence is around 2.35 eV, and that zinc vacancies are responsible for the emission, but in ZnO films with an excess of zinc, the maximum of luminescence is around 2.53 eV, and in this case, oxygen vacancies acting as deep acceptor centers are responsible for the observed emission.

In our case, the DLE emission observed was centered at 2.53 eV, and according to the results presented above, the films contained both zinc in excess and oxygen vacancies; therefore, those defects could be involved in the PL phenomena.

As can be observed in Figure 12, all DLE emissions obtained in these films were centered at 2.50 eV (495 nm), and except for curve f, all films presented a broadening from 414 nm to 630 nm, and as mentioned above, the most intense peak could be attributed mainly to oxygen vacancies, and the broadening of curve could be assigned to a combination of different deep levels, either intrinsic or extrinsic.

Tantalum atoms introduced shallow donors in ZnO and modified its electrical conductivity, but in this case, those centers did not seem to play any effect on the PL phenomena, because zinc interstitials and oxygen vacancies are the defects involved in the PL phenomena.

Because the effect of the introduction of Ta atoms in ZnO is to increase the ratio of the volume to surface area as the structures area become smaller, the increase on the PL intensity could be attributed to a variation in the surface area of these films, as has been shown in the SEM section.

3.5. Hall Measurements

It is well-established that ZnO has natural type-n conductivity because of the formation of intrinsic defects, such as oxygen vacancies and zinc interstitials, during the synthesis. Oxygen vacancies can be easily formed and can contribute with two free electrons to the conduction band. However, it has been shown that those defects are actually deep donors, and they cannot contribute to the n-type conductivity. Therefore, zinc interstitials seem like the most probable defect acting as a donor in undoped ZnO films.

Carrier concentration, mobility, and the resistivity of films measured by the Hall–Vander Paw technique, are presented in Figure 13. The majority carrier mobility first increases up to a maximum value, and afterwards it reduces, as the Ta_2O_5 percentage is increased in the growth, while the carrier concentration increases as the Ta_2O_5 percentage increases.

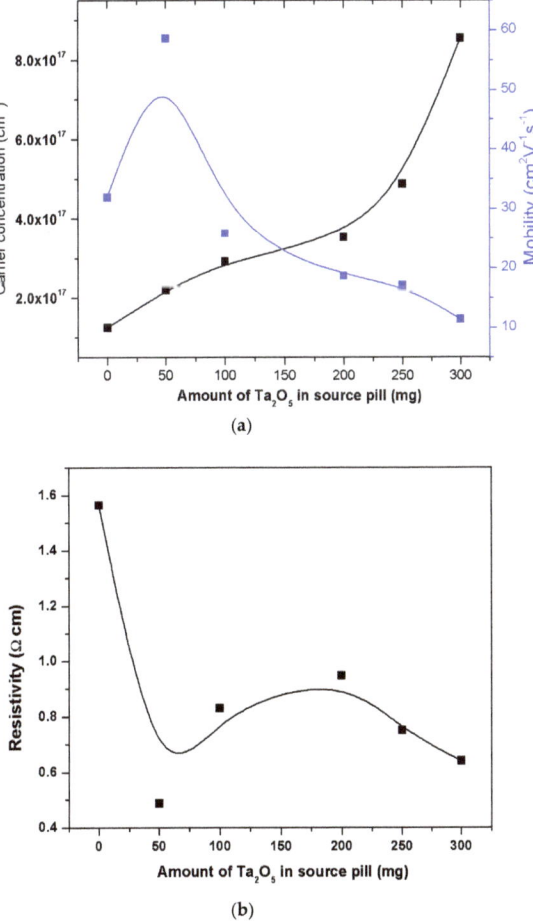

Figure 13. Electrical characterization of ZnO- and ZnO:Ta-doped films: (**a**) carrier concentration and mobility, and (**b**) resistivity.

The minimum value of the carrier concentration is for undoped samples (Figure 13a); therefore, the increase of the concentration was for the incorporation of the Ta atoms in the Zn positions of ZnO, and they contributed with electrons to the conduction band. The carrier concentration increased as the amount of Ta_2O_5 was incremented in the source pellet; this could be a clear reason for the increment in carrier concentration in ZnO:Ta-doped films, since if a tantalum atom replaces a zinc atom, this tantalum could have an oxidation state of $Ta1^+$, $Ta2^+$, or $Ta3^+$ with regards to Zn; this means that tantalum is capable of contributing with free electrons.

The mobility (Figure 13a) initially increased, with Ta_2O_5 reaching a maximum value in the film fabricated with 50 mg of Ta_2O_5 in the source; after that, it started to decrease. This effect has also been observed by Subha et al. [57,58], and is attributed to an excess of carriers in the ZnO lattice. A small increment observed in mobility, was owing to a high number of free carriers that were generated by tantalum atoms, and the relatively greater crystal size of the shell. Nevertheless, this mobility was reduced as the carrier concentration increased, because this carrier created a high number of collisions between them, besides the increase in the number of grain boundaries, due to a reduction in the crystal size in the shell. There are reports of ZnO with other dopants, in which it has been observed that foreign atoms inside the lattice of ZnO could act as charge trapping sites [33,42], which prevents the mobility of these free chargers. In our case, Ta atoms could be creating a similar effect, and therefore, the mobility in our films decreases with the amount of Ta_2O_5 in the source pellet.

With regard to the resistivity of the films, it can be appreciated in Figure 13b that initially for ZnO film, this had the highest value, but when these were doped with tantalum atoms, resistivity was around 0.5–1.0 Ωcm, but this tended to decrease, and this effect was attributed to an excess of carriers generated by donors from Ta atoms incorporated in the ZnO lattice.

For the conduction process in these type of films and to attempt to explain the results above the following, a model is proposed in Figure 14.

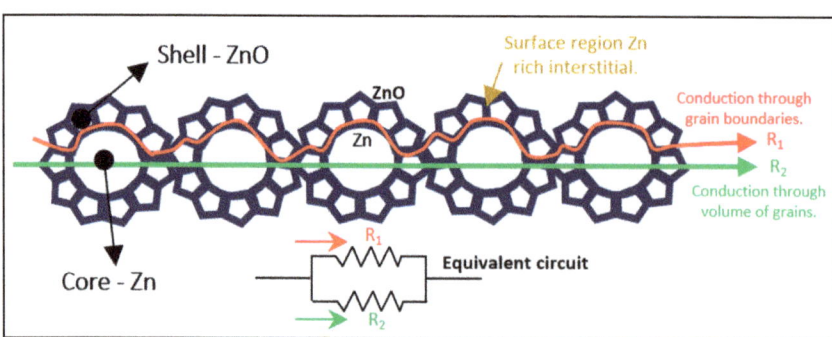

Figure 14. A model of conductivity through the crystal in the film. Two possible paths for conduction: intergrain (R_1) and through the volume of the grains (R_2).

First of all, we consider that the conduction of carriers in core shell structures could be a bit different from the pure crystals. As seen in the SEM pictures (Figure 9), the core shell structures were composed by a metallic core, and a shell formed by the oxide crystals. Due to the high temperature of the processes (above of the melting point of the zinc ~420 °C), the zinc atoms diffused through the grain boundaries to release the internal pressure, creating zinc-rich regions between grains. These zinc atoms became suboxides with a higher concentration of zinc interstitials at the surface than in the grain volume. The effect created a surface path for electric conduction, while the grain volume with lower concentrations of interstitials had a low contribution to the conductivity.

When tantalum was introduced into the ZnO matrix, the concentration of the carriers in the volume increased, and the grain was capable of being a transport carrier, and the resistivity of the film decreased, as shown in Figure 13.

4. Conclusions

ZnO- and ZnO:Ta-doped films were obtained through a HFCVD approach; the films were been obtained at 800 °C under a hydrogen atmosphere; the pellet source was fabricated with a powder mixture of ZnO and Ta_2O_5. The films obtained were studied for their structural, morphological, compositional, optical, and electrical properties; Ta atoms were incorporated into the ZnO wurtzite structure but they did not affect this crystal lattice, owing to the substitution of the Zn atoms with the Ta atoms, and these possess similar atomic radii; nevertheless, the Ta atoms tended to the growth direction (201) of ZnO. Raman spectroscopy confirms a majority contribution of ZnO in lattice, but Ta was not observed by this technique. According to EDS, from the films obtained, both ZnO- and ZnO:Ta-doped films are Zn-rich, but tantalum is present at a low concentration, though tantalum plays an important role in the morphological structure, and the roughness decreases as the amount of Ta is increased in the pellet source, and on the other hand, the density of the big sphere shapes tended to disappear; moreover, this technique confirms structures of a core–shell type. Photoluminescence studies reveal that the films obtained emit in a range of the green visible spectra, this effect mainly being due to oxygen vacancies; the tantalum atoms did not seem to play an important role in the creation of new centers of generation/recombination, but only in the intensity of emission observed in 495 nm; this was owing to the increment in the surface area of the films. Electrical measurements show that the films obtained are n-type, and that the tantalum atoms incorporated into the lattice of ZnO are contributing with a donor, increasing the carrier concentration as Ta is increased in the film; the resistivity and the sheet resistance also decrease when there is a greater amount of tantalum in the film.

Author Contributions: V.H. and T.D.-B. wrote, conceived and designed the experiments, E.R.-C., G.G.-S., and R.G. provide resources and systems to work, C.M., A.C., E.R., R.R., and F.G.N.-C. provide support in review and editing this work.

Acknowledgments: Authors gratefully acknowledge N. Rutilo Silva from IFUAP for FESEM images supported in this work, and M. Aceves from INAOE for PL measurements, as well as BUAP financial for support in publication fee, and CONACyT for Ph.D. support 304754 in the Institute of semiconductor devices program from BUAP.

Conflicts of Interest: The authors declare no conflict of interest. The funders had no role in the design of the study; in the collection, analyses, or interpretation of data; in the writing of the manuscript, and in the decision to publish the results.

References

1. Versteegh, M.A.M.; Vanmaekelbergh, D.; Dijkhuis, J.I. Room-Temperature Laser Emission of ZnO Nanowires Explained by Many-Body Theory. *Phys. Rev. Lett.* **2012**, *108*, 157402. [CrossRef] [PubMed]
2. Willander, M.; Nur, O.; Zhao, Q.X.; Yang, L.L.; Lorenz, M.; Cao, B.Q.; Ziga Pérez, J.; Czekalla, C.; Zimmermann, G.; Grundmann, M.; et al. Zinc oxide nanorod based photonic devices: Recent progress in growth, lightemitting diodes and lasers. *Nanotechnology* **2009**, *20*, 332001. [CrossRef] [PubMed]
3. Willander, M.; Nur, O.; Sadaf, J.R.; Qadir, M.I.; Zaman, S.; Zainelabdin, A.; Bano, N.; Hussain, I. Luminescence from zinc oxide nanostructures and polymers and their hybrid devices. *Materials* **2010**, *3*, 2643–2667. [CrossRef]
4. Guo, L.; Zhang, H.; Zhao, D.; Li, B.; Zhang, Z.; Jiang, M.; Shen, D. High responsivity ZnO nanowires based UV detector fabricated by the dielectrophoresis method. *Sens. Actuators B Chem.* **2012**, *166–167*, 12–16. [CrossRef]
5. Könenkamp, R.; Nadarajah, A.; Word, R.C.; Meiss, J.; Engelhardt, R. ZnO nanowires for LED and field-emission displays. *J. Soc. Inf. Disp.* **2008**, *16*, 609–613. [CrossRef]
6. Zheng, K.; Shen, H.; Li, J.; Sun, D.; Chen, G.; Hou, K.; Li, C.; Lei, W. The fabrication and properties of field emission display based on ZnO tetrapod-liked nanostructure. *Vacuum* **2008**, *83*, 261–264. [CrossRef]
7. Seelig, E.W.; Tang, B.; Yamilov, A.; Cao, H.; Chang, R.P.H. Self-assembeled 3D photonic crystals from ZnO colloidal spheres. *Mater. Chem. Phys.* **2002**, *9712*, 1–7.

8. Pietruszka, R.; Witkowski, B.S.; Gieraltowska, S.; Caban, P.; Wachnicki, L.; Zielony, E.; Gwozdz, K.; Bieganski, P.; Placzek-Popko, E.; Godlewski, M. New efficient solar cell structures based on zinc oxide nanorods. *Sol. Energy Mater. Sol. Cells* **2015**, *143*, 99–104. [CrossRef]
9. Vittal, R.; Ho, K.C. Zinc oxide based dye-sensitized solar cells: A review. *Renew. Sustain. Energy Rev.* **2017**, *70*, 920–935. [CrossRef]
10. Manthina, V.; Agrios, A.G. Band edge engineering of composite photoanodes for dye-sensitized solar cells. *Electrochim. Acta* **2015**, *169*, 416–423. [CrossRef]
11. Yazdi, M.A.P.; Martin, N.; Monsifrot, E.; Briois, P.; Billard, A. ZnO nano-tree active layer as heavy hydrocarbon sensor: From material synthesis to electrical and gas sensing properties. *Thin Solid Films* **2015**, *596*, 128–134. [CrossRef]
12. Chaudhary, S.; Umar, A.; Bhasin, K.K.; Baskoutas, S. Chemical sensing applications of ZnO nanomaterials. *Materials* **2018**, *11*, 287. [CrossRef] [PubMed]
13. Manthina, V.; Agrios, A.G. Single-pot ZnO nanostructure synthesis by chemical bath deposition and their applications. *Nano-Struct. Nano-Objects* **2016**, *7*, 1–11. [CrossRef]
14. Yang, P.F.; Wen, H.C.; Jian, S.R.; Lai, Y.S.; Wu, S.; Chen, R.S. Characteristics of ZnO thin films prepared by radio frequency magnetron sputtering. *Microelectron. Reliab.* **2008**, *48*, 389–394. [CrossRef]
15. Villanueva, Y.Y.; Liu, D.R.; Cheng, P.T. Pulsed laser deposition of zinc oxide. *Thin Solid Films* **2006**, *501*, 366–369. [CrossRef]
16. Lehraki, N.; Aida, M.S.; Abed, S.; Attaf, N.; Attaf, A.; Poulain, M. ZnO thin films deposition by spray pyrolysis: Influence of precursor solution properties. *Curr. Appl. Phys.* **2012**, *12*, 1283–1287. [CrossRef]
17. Wallace, R.; Brown, A.P.; Brydson, R.; Wegner, K.; Milne, S.J. Synthesis of ZnO nanoparticles by flame spray pyrolysis and characterisation protocol. *J. Mater. Sci.* **2013**, *48*, 6393–6403. [CrossRef]
18. Ohmagari, S.; Matsumoto, T.; Umezawa, H.; Mokuno, Y. Ohmic contact formation to heavily boron-doped p+ diamond prepared by hot-filament chemical vapor deposition. *MRS Adv.* **2016**, *1*, 3489–3495. [CrossRef]
19. Deshpande, S.; Dupuie, J.; Gulari, E. Filament-activated chemical vapour deposition of nitride thin films. *Adv. Mater. Opt. Electron.* **1996**, *6*, 135–146. [CrossRef]
20. Deshpande, S.V.; Dupuie, J.L.; Gualari, E. Hot filament assisted deposition of silicon nitride thin films. *Appl. Phys. Lett.* **1992**, *61*, 1420–1422. [CrossRef]
21. Silva-Andrade, F.; Chávez, F.; Gómez, E. Epitaxial GaAs growth using atomic hydrogen as the reactant. *J. Appl. Phys.* **1994**, *76*, 1946–1947. [CrossRef]
22. López, J.A.L.; López, J.C.; Valerdi, D.E.V.; Salgado, G.G.; Díaz-Becerril, T.; Pedraza, A.P.; Gracia, F.J.F. Morphological, compositional, structural, and optical properties of Si-nc embedded in SiOx films. *Nanoscale Res. Lett.* **2012**, *7*, 604. [CrossRef] [PubMed]
23. Mao, H.Y.; Lo, S.Y.; Wuu, D.S.; Wu, B.R.; Ou, S.L.; Hsieh, H.Y.; Horng, R.H. Hot-wire chemical vapor deposition and characterization of p-type nanocrystalline Si films for thin film photovoltaic applications. *Thin Solid Films* **2012**, *520*, 5200–5205. [CrossRef]
24. Mendoza, F.; Limbu, T.B.; Weiner, B.R.; Morell, G. Large-area bilayer graphene synthesis in the hot filament chemical vapor deposition reactor. *Diam. Relat. Mater.* **2015**, *51*, 34–38. [CrossRef]
25. Wang, B.B.; Zhu, M.K.; Ostrikov, K.; Shao, R.W.; Zheng, K. Structure and photoluminescence of molybdenum selenide nanomaterials grown by hot filament chemical vapor deposition. *J. Alloys Compd.* **2015**, *647*, 734–739. [CrossRef]
26. Mortazavi, S.H.; Ghoranneviss, M.; Dadashbaba, M.; Alipour, R. Synthesis and investigation of silicon carbide nanowires by HFCVD method. *Bull. Mater. Sci.* **2016**, *39*, 953–960. [CrossRef]
27. Ramos, J.R.; Morales, C.; García, G.; Díaz, T.; Rosendo, E.; Santoyo, J.; Oliva, A.I.; Galeazzi, R. Optical and structural analysis of ZnS core-shell type nanowires. *J. Alloys Compd.* **2018**, *736*, 93–98. [CrossRef]
28. Janotti, A.; Van De Walle, C.G. Native point defects in ZnO. *Phys. Rev. B Condens. Matter Mater. Phys.* **2007**, *76*, 165202. [CrossRef]
29. Jayakumar, O.D.; Sudarsan, V.; Sudakar, C.; Naik, R.; Vatsa, R.K.; Tyagi, A.K. Green emission from ZnO nanorods: Role of defects and morphology. *Scr. Mater.* **2010**, *62*, 662–665. [CrossRef]
30. Gahlaut, U.P.S.; Kumar, V.; Pandey, R.K.; Goswami, Y.C. Highly luminescent ultra small Cu doped ZnO nanostructures grown by ultrasonicated sol-gel route. *Optik* **2016**, *127*, 4292–4295. [CrossRef]

31. Wang, Y.; Liu, N.; Chen, Y.; Yang, C.; Liu, W.; Su, J.; Li, L.; Gao, Y. Multicolour electroluminescence from light emitting diode based on ZnO:Cu/p-GaN heterojunction at positive and reverse bias voltage. *RSC Adv.* **2015**, *5*, 104386–104391. [CrossRef]
32. Muthukumaran, S.; Gopalakrishnan, R. Structural, FTIR and photoluminescence studies of Cu doped ZnO nanopowders by co-precipitation method. *Opt. Mater.* **2012**, *34*, 1946–1953. [CrossRef]
33. Klingshirn, C. ZnO: From basics towards applications. *Phys. Status Solidi Basic Res.* **2007**, *244*, 3027–3073. [CrossRef]
34. Özgür, Ü.; Alivov, Y.I.; Liu, C.; Teke, A.; Reshchikov, M.A.; Doğan, S.; Avrutin, V.; Cho, S.J.; Morkoç, H. A comprehensive review of ZnO materials and devices. *J. Appl. Phys.* **2005**, *98*, 1–103. [CrossRef]
35. Chitra, M.; Uthayarani, K.; Rajasekaran, N.; Girija, E.K. Preparation and characterisation of Al doped ZnO nanopowders. *Phys. Procedia* **2013**, *49*, 177–182. [CrossRef]
36. Liu, M.; Kitai, A.H.; Mascher, P. Point defects and luminescence centers in zinc oxide and zinc oxide doped with manganese. *J. Lumin.* **1992**, *54*, 35–42. [CrossRef]
37. López, R.; Díaz, T.; García, G.; Rosendo, E.; Galeazzi, R.; Coyopol, A.; Juárez, H.; Pacio, M.; Morales, F.; Oliva, A.I. Fast formation of surface Oxidized Zn Nanorods and urchin-like microclusters. *Adv. Mater. Sci. Eng.* **2014**, *2014*. [CrossRef]
38. Richard, D.; Romero, M.; Faccio, R. Experimental and theoretical study on the structural, electrical and optical properties of tantalum-doped ZnO nanoparticles prepared via sol-gel acetate route. *Ceram. Int.* **2018**, *44*, 703–711. [CrossRef]
39. Yuan, L.; Wang, C.; Cai, R.; Wang, Y.; Zhou, G. Temperature-dependent growth mechanism and microstructure of ZnO nanostructures grown from the thermal oxidation of zinc. *J. Cryst. Growth* **2014**, *390*, 101–108. [CrossRef]
40. Cheng, Y.; Cao, L.; He, G.; Yao, G.; Song, X.; Sun, Z. Preparation, microstructure and photoelectrical properties of Tantalum-doped zinc oxide transparent conducting films. *J. Alloys Compd.* **2014**, *608*, 85–89. [CrossRef]
41. Krishnan, R.R.; Vinodkumar, R.; Rajan, G.; Gopchandran, K.G.; Mahadevan Pillai, V.P. Structural, optical, and morphological properties of laser ablated ZnO doped Ta_2O_5 films. *Mater. Sci. Eng. B Solid-State Mater. Adv. Technol.* **2010**, *174*, 150–158. [CrossRef]
42. Bang, K.; Son, G.C.; Son, M.; Jun, J.H.; An, H.; Baik, K.H.; Myoung, J.M.; Ham, M.H. Effects of Li doping on the structural and electrical properties of solution-processed ZnO films for high-performance thin-film transistors. *J. Alloys Compd.* **2018**, *739*, 41–46. [CrossRef]
43. Khan, A. Raman Spectroscopic Study of the ZnO Nanostructures. *J. Pak. Mater. Soc.* **2010**, *4*, 5–9.
44. Schumm, M. ZnO-Based Semiconductors Studied by Raman Spectroscopy: Semimagnetic Alloying, Doping, and Nanostructures. Ph.D. Thesis, Julius–Maximilians University, Würzburg, Germany, July 2008.
45. Soosen, S.M.; Koshy, J.; Chandran, A.; George, K.C. Optical phonon confinement in ZnO nanorods and nanotubes. *Indian J. Pure Appl. Phys.* **2010**, *48*, 703–708.
46. Zhang, R.; Yin, P.G.; Wang, N.; Guo, L. Photoluminescence and Raman scattering of ZnO nanorods. *Solid State Sci.* **2009**, *11*, 865–869. [CrossRef]
47. Tzolov, M.; Tzenov, N.; Dimova-Malinovska, D.; Kalitzova, M.; Pizzuto, C.; Vitali, G.; Zollo, G.; Ivanov, I. Vibrational properties and structure of undoped and Al-doped ZnO films deposited by RF magnetron sputtering. *Thin Solid Films* **2000**, *379*, 28–36. [CrossRef]
48. Goff, A.H.-L.; Joiret, S.; Saïdani, B.; Wiart, R. In-situ Raman spectroscopy applied to the study of the deposition and passivation of zinc in alkaline electrolytes. *J. Electroanal. Chem.* **1989**, *263*, 127–135. [CrossRef]
49. Marchebois, H.; Joiret, S.; Savall, C.; Bernard, J.; Touzain, S. Characterization of zinc-rich powder coatings by EIS and Raman spectroscopy. *Surf. Coat. Technol.* **2002**, *157*, 151–161. [CrossRef]
50. Janotti, A.; Van De Walle, C.G. Fundamentals of zinc oxide as a semiconductor. *Rep. Prog. Phys.* **2009**, *72*, 126501. [CrossRef]
51. Van De Walle, C.G.; Neugebauer, J. First-principles calculations for defects and impurities: Applications to III-nitrides. *J. Appl. Phys.* **2004**, *95*, 3851–3879. [CrossRef]
52. López, R.; García, G.; Díaz, T.; Coyopol, A.; Rosendo, E.; Galeazzi, R.; Juárez, H.; Pacio, M. Low temperature growth of Zn-ZnO microspheres by atomic hydrogen assisted-HFCVD. *IOP Conf. Ser. Mater. Sci. Eng.* **2013**, *45*, 012016. [CrossRef]

53. López, R.; Díaz, T.; García, G.; Galeazzi, R.; Rosendo, E.; Coyopol, A.; Pacio, M.; Juárez, H.; Oliva, A.I. Structural properties of Zn-ZnO core-shell microspheres grown by hot-filament CVD technique. *J. Nanomater.* **2012**, *2012*, 865321. [CrossRef]
54. Lin, J.H.; Patil, R.A.; Devan, R.S.; Liu, Z.A.; Wang, Y.P.; Ho, C.H.; Liou, Y.; Ma, Y.R. Photoluminescence mechanisms of metallic Zn nanospheres, semiconducting ZnO nanoballoons, and metal-semiconductor Zn/ZnO nanospheres. *Sci. Rep.* **2014**, *4*, 6967. [CrossRef] [PubMed]
55. Zhao, C.X.; Li, Y.F.; Zhou, J.; Li, L.Y.; Deng, S.Z.; Xu, N.S.; Chen, J. Large-scale synthesis of bicrystalline ZnO nanowire arrays by thermal oxidation of zinc film: Growth mechanism and high-performance field emission. *Cryst. Growth Des.* **2013**, *13*, 2897–2905. [CrossRef]
56. Wu, Y.; Li, C.; Li, M.; Li, H.; Xu, S.; Wu, X.; Yang, B. Microstructural and optical properties of Ta-doped ZnO films prepared by radio frequency magnetron sputtering. *Ceram. Int.* **2016**, *42*, 10847–10853. [CrossRef]
57. Ravichandran, K.; Subha, K.; Dineshbabu, N.; Manivasaham, A. Enhancing the electrical parameters of ZnO films deposited using a low-cost chemical spray technique through Ta doping. *J. Alloys Compd.* **2016**, *656*, 332–338. [CrossRef]
58. Subha, K.; Ravichandran, K.; Sriram, S. Combined influence of fluorine doping and vacuum annealing on the electrical properties of ZnO:Ta films. *Appl. Surf. Sci.* **2017**, *409*, 413–425. [CrossRef]
59. Li, G.; Kawi, S. High-surface-area SnO: A novel semiconductor-oxide. *Mater. Lett.* **1998**, *34*, 99–102. [CrossRef]
60. Bain, L.E.; Collazo, R.; Hsu, S.H.; Latham, N.P.; Manfra, M.J.; Ivanisevic, A. Surface topography and chemistry shape cellular behavior on wide band-gap semiconductors. *Acta Biomater.* **2014**, *10*, 2455–2462. [CrossRef] [PubMed]
61. Soni, U.; Sapra, S. The Importance of Surface in Core—Shell Semiconductor Nanocrystals. *J. Phys. Chem.* **2010**, *114*, 22514–22518. [CrossRef]
62. Rodnyi, P.A.; Khodyuk, I.V. Optical and luminescence properties of zinc oxide (Review). *Opt. Spectrosc.* **2011**, *111*, 776–785. [CrossRef]
63. Studenikin, S.A.; Golego, N.; Cocivera, M. Fabrication of green and orange photoluminescent, undoped ZnO films using spray pyrolysis. *J. Appl. Phys.* **1998**, *84*, 2287–2294. [CrossRef]
64. Vanheusden, K.; Warren, W.L.; Seager, C.H.; Tallant, D.R.; Voigt, J.A.; Gnade, B.E. Mechanisms behind green photoluminescence in ZnO phosphor powders. *J. Appl. Phys.* **1996**, *79*, 7983–7990. [CrossRef]

© 2018 by the authors. Licensee MDPI, Basel, Switzerland. This article is an open access article distributed under the terms and conditions of the Creative Commons Attribution (CC BY) license (http://creativecommons.org/licenses/by/4.0/).

Article

Simulation and Analysis of Single-Mode Microring Resonators in Lithium Niobate Thin Films

Huangpu Han [1,2], Bingxi Xiang [3,4,*] and Jiali Zhang [5]

[1] College of Electric and Electronic Engineering, Zibo Vocational Institute, Zibo 255314, China; pupuhan@126.com
[2] School of Physics, Shandong University, Jinan 250100, China
[3] College of New Materials and New Energies, Shenzhen Technology University, Shenzhen 518118, China
[4] Shenzhen Key Laboratory of Laser Engineering, College of Optoelectronic Engineering, Shenzhen University, Shenzhen 518060, China
[5] College of General Education, Weifang University of Science and Technology, Weifang 262700, China; jialigg@126.com
* Correspondence: xiangbingxi@sztu.edu.cn; Tel.: +86-135-3085-0236

Received: 4 August 2018; Accepted: 22 August 2018; Published: 24 August 2018

Abstract: The single-mode microring resonators on lithium niobate thin films were designed and simulated using 2.5-D variational finite difference time domain mode simulations from Lumerical mode Solutions. The single-mode conditions and the propagation losses of lithium niobate planar waveguide with different SiO_2 cladding layer thicknesses were studied and compared systematically. The optimization of design parameters such as radii of microrings and gap sizes between channel and ring waveguides were determined. The key issues affecting the resonator design such as free spectral range and Quality Factor were discussed. The microring resonators had radius R = 20 µm, and their transmission spectrum had been tuned using the electro-optical effect.

Keywords: microring resonator; integrated photonic; varfdtd; electro-optical; LNOI

1. Introduction

Lithium niobate ($LiNO_3$, LN) is one of the most promising materials because of its excellent electro-optical, piezoelectric, pyroelectric, photo-elastic and non-linear properties [1]. Due to the high electro-optical coefficient (γ_{33} = 31.2 pm/V) in LN, highly efficient electro-optical modulator in such material is very promising and always an interesting topic in optical interconnect technology [2]. In last ten years, high-refractive-index contrast, which is LN thin film on a low refractive index SiO_2 cladding layer or other substrates (lithium niobate on insulator, LNOI) has emerged, as it is compatible with the silicon on insulator (SOI) manufacturing technology [3]. LNOI is an interesting platform for integrated photonics due to the good confinement and strong guiding of light [4].

The microring resonator is one of the important elements in integrated photonic systems. Optical microring resonators are the promising candidates for a variety of applications, including wavelength filtering, multiplexing, switching and modulation [5–7]. The basic structure of the microring resonator consists of a channel and a ring. The microring resonators have been designed on various types of materials such as silicon [8], germanium-doped silica-on-silicon [9], polymers [10], lithium niobate [11–13] and so forth. The utmost waveguide width with the single-mode behavior is the optimum condition, because no higher order modes were excited and the fundamental mode had the highest light confinement [14]. Therefore, the single-mode condition in the waveguide should be fulfilled in order to prevent signal distortion during transmission. Many articles have reported the fabrication of microring resonators in LN on insulator [11–13] and simulation of microring resonators in SOI [15,16], but few reports have involved the simulation of single-mode microring resonators in LNOI in the literature so far. Therefore, a search for the simulation and analysis of single-mode microring resonators in LNOI seems to be important. Our work has focused on LNOI-based devices in order to obtain a high Quality Factor (Q-factor) and wide free spectral range (FSR) that are crucial in developing ultra-compact integrations of planar lightwave circuits [4,17,18].

In this paper, we presented results from finite difference time domain (FDTD) numerical simulations. Key optical design parameters of laterally coupled LNOI-based microring resonators were characterized using 2.5-D variational FDTD (varFDTD). Waveguide simulations based on full-vectorial finite difference method were performed by varying the geometrical parameters of waveguide configurations to investigate the single-mode conditions and the propagation losses of LN planar waveguides at different SiO_2 cladding layer thicknesses. In order to obtain an improved wavelength filtering, we performed the optimization for the design parameters which includes ring radii and gap sizes. These parameters affected FSR and Q-factor directly. As a very important element for practical application we described the electro-optical tunable microring resonators on LNOI.

2. Device Description

The material of the device studied in this paper was a z-cut LN thin film stuck to SiO_2 cladding layer deposited on a LN substrate. The schematic of a microring resonator is shown in Figure 1. The modulator consisted of a ring resonator made by the z-cut LN thin film, which was coupled to a channel. If the signal was on-resonance with the ring, then that signal coupled into the cavity from channel. The device parameters such as ring radii and gap sizes between channel and ring resonator were optimized to have the optimum characteristics.

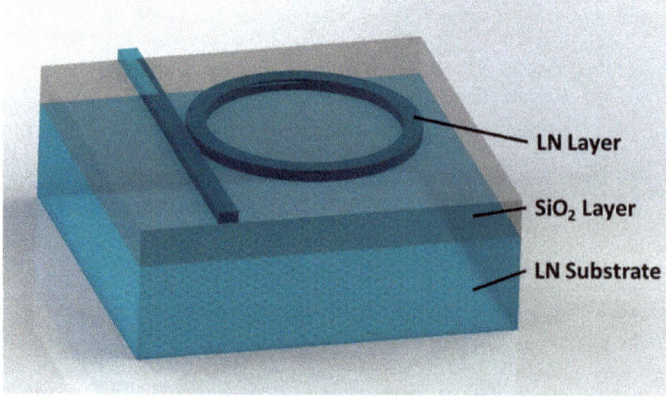

Figure 1. A schematic of the waveguide-coupled microring resonator on LNOI.

The varFDTD method with perfectly matched layers boundary conditions (PML) in this work could show how to design and simulate a microring resonator using the finite difference eigen (FDE) mode solver to achieve a desired FSR and Q factor for a LNOI based. The varFDTD method was a direct time and space solution for solving Maxwell's equations in complex geometries. By performing Fourier transforms, such as the Poynting vector, normalized transmission, and far field projections could be obtained. PML boundaries could absorb electromagnetic energy incident upon them, allowing radiation to propagate out of the computational area without interfering with the field inside [19]. The total ring loss was due to the bend loss and other scattering losses at the coupling region. For the high index contrast wave-guides we were considering, at wavelengths around 1.55 µm, the propagation loss and bend loss were small. For this analysis we would neglect all losses, but the different loss contributions could easily be considered in a more detailed analysis [20].

3. Results and Discussion

The high refractive index difference between the LN (n_o = 2.2129 and n_e = 2.1395 at λ = 1.55 µm) [21] and the SiO$_2$ (n = 1.46) required the waveguide thickness and the waveguide width to be smaller than a limiting value to achieve single-mode operation. The single mode conditions had been firstly simulated.

We had calculated the modal curves at λ = 1.55 µm for LN ridge waveguides with 0.7 µm width on SiO$_2$ layer. The effective index dependence on the thickness is presented in Figure 2. The first order mode of the transverse electric (TE) and transverse magnetic (TM) modes appeared at the LN thicknesses of 0.86 µm and 0.84 µm, respectively. To ensure that only one electric field intensity peak was supported in the vertical direction of the LN thin film, the thickness of LN thin film should be less than this critical value. In the following simulation, the thickness of LN thin film was all selected as 0.5 µm.

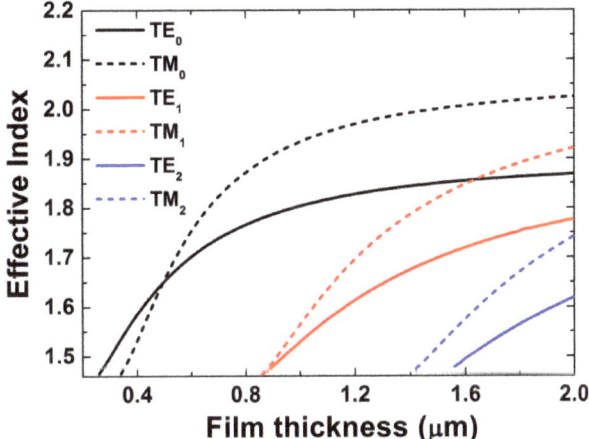

Figure 2. Effective index of the TE (solid lines) and TM (dashed lines) modes in LN waveguides as a function of the film thickness for a waveguide with 0.7 µm width; the modes were calculated at λ = 1.55 µm.

The dependence of the effective index on the waveguide width, assuming a 0.5-µm thick waveguide is shown in Figure 3. The curves showed the boundary between the single-mode and multi-mode conditions. For LN ridge waveguides, the single mode conditions of the TE and TM modes were 1.08 µm and 1.04 µm, respectively. In the following simulations, the widths of LN ridges were all selected as 0.7 µm to ensure that only one mode was supported in the LN thin film.

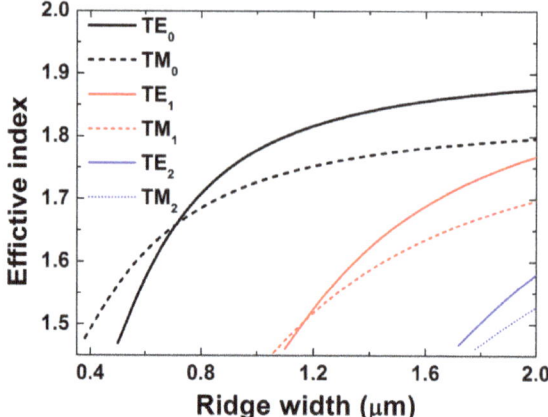

Figure 3. Effective index of the TE (solid lines) and TM (dashed lines) modes in LN waveguides as a function of the width for a 0.5 µm thick film. The modes were calculated at λ = 1.55 µm.

The SiO$_2$ layer worked as the cladding layer with a thickness of a limiting value, which was sufficiently thick to prevent the field penetration into the bottom layer. The propagation losses of LN planar waveguide with the different SiO$_2$ layer thicknesses are shown in Figure 4. The propagation losses descended with increasing SiO$_2$ cladding layer thicknesses. At the same SiO$_2$ layer thickness, the propagation losses of TE modes were less than those of the TM modes. The diffracted field radiated in the transmission of light, some power was lost into the substrate, but the structure of LN-SiO$_2$-Au had a metal bottom reflector, the power transmitted to the substrate was reflected by the Au bottom reflector, which resulted a decrease of the propagation losses [22]. The propagation losses of LN-SiO$_2$-Au were less than those of LN-SiO$_2$-LN at the same SiO$_2$ layer thickness. At a SiO$_2$ layer thickness of 2 µm, the LN planar waveguide losses were all less than 10^{-3} dB/cm, which is negligible. Therefore, the thicknesses of SiO$_2$ layer were all selected as 2 µm in the following simulation.

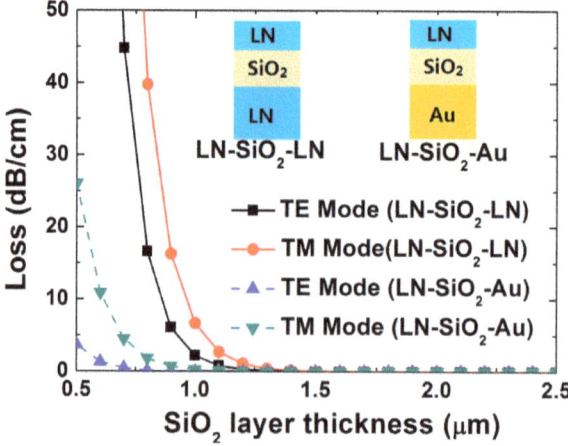

Figure 4. The propagation losses of LN planar waveguide with the different SiO$_2$ layer thicknesses.

A critical dimension in this optical structure was the gap separating the ring from the tangential waveguide. The gap size determined the input and output couple ratios of the microring resonator, which determined the magnitude of the finesse and the at-resonance transmittance in turn. In the case of a microring coupled to channel, the gaps were very small due to the strong optical confinement and the small coupling interaction length [14].

Because the FSR defined as the separation between two adjacent resonant wavelengths, increased and might become even larger than the wavelength range used for wavelength division multiplex (WDM) applications. However, one of the most critical factors limiting the minimum useful ring radius was the bending loss. These could be qualitatively understood by describing the bend as a straight waveguide, while the effective index was decreasing function in the radial direction. This implied that at a certain distance from the waveguide core, the solution of the Maxwell equations became a radiating field; this radiation was a loss source, as in a leaky waveguide [11].

Using varFDTD, we have calculated Q-factor, bending loss and FSR for a single-mode waveguide at different ring radii and gap sizes (channel and ring width w = 0.7 μm, λ = around 1.55 μm). The results are shown in Figures 5 and 6. According to the Figure 5a, when the radius was smaller than 20 μm, the Q-factor significantly ascended with increased radii, while it was larger than 20 μm, it preserved almost unchangeable values. We also observed that Q-factor ascended with increasing gap sizes and the TE and TM modes showed different results in terms of the achievable Q-factor for this geometry. In the case of gap = 0.29 μm, a Q-factor more than 10,000 was achieved for the TE mode whilst for the TM mode the Q-factor was significantly smaller ~2000. Figure 5b shows Bending loss descended with increasing ring radius, and the bending losses are very low as the radii more than 20 μm. Loss in practical structures would inevitably come up for the following two reasons: on one hand, the scattering by the residual roughness of the etched walls of the photonic wires [17], on the other hand, the modal mismatch at the interface between SiO_2 layer and LN thin film [23]. However, the losses of the LN waveguides could be reduced as much as possible by optimizing the fabrication process of waveguides. At 1590 nm wavelength, the propagation losses of sub-wavelength scale LN waveguides could be as low as 2.7 dB/m by Ar^+ ion etching [12]. As shown in Figure 6, FSR descended with increasing ring radius. Both TE and TM modes achieved the largest FSR for the ring with the smallest radius 3 μm. When the radius was larger than 40 μm and the gap size was 0.1 μm, the FWHM (full-width at half-maximum) of the dip was close to the FSR, the dip became less obvious. Because the effective index of TE mode was larger than that of TM mode, the Q-factor of TE mode was greater than that of TM mode, and FSR of TE mode was less than that of TM mode.

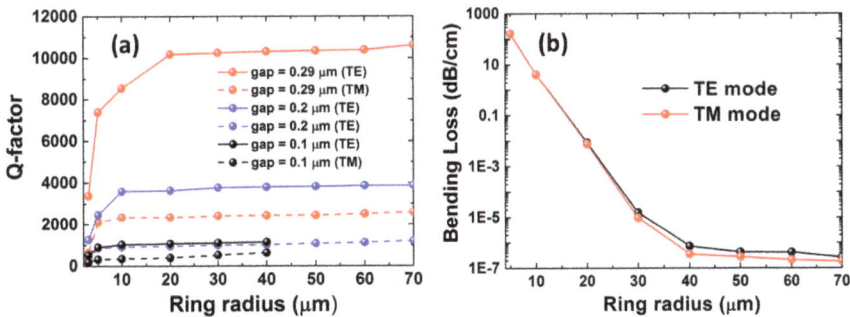

Figure 5. (a) Q-factor of microring resonator as different ring radii for different gap sizes, (b) Bending loss variation as bending radius. The modes were calculated at λ = around 1.55 μm.

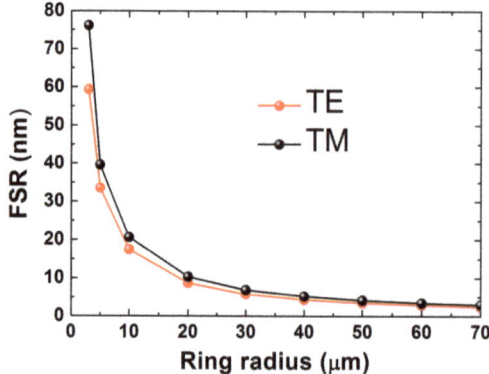

Figure 6. FSR of microring resonators as different ring radii. The modes were calculated at λ = around 1.55 μm.

The electro-optical properties of LN microring resonator have been simulated by applying a static electric field to the device electrodes to shift transmission spectrum. In Figure 7, a LN microring resonator is embedded in the middle of SiO$_2$ layer, and the electrodes are placed over below and above the SiO$_2$ layer. A tuning range of microring resonators depend on the strength of the electro-optical effect. However, the electric field achieved in the LN film was relatively weak because the upper and lower cladding materials exhibited one order of magnitude smaller dielectric constants. Due to the large difference between the dielectric constant of the LN thin film (ε_{LN} = 28.5) and the SiO$_2$ (ε_{SiO_2} = 3.9), the field in the LN thin film was considerably far less than in the SiO$_2$ layer. The field in the two layers could be calculated from the applied voltage ΔV using the continuity of the vertical component of the electric displacement field D:

$$E_{LN}\, d_{LN} + E_{SiO_2}\, d_{SiO_2} = \Delta V, \tag{1}$$

$$D = \varepsilon_0\, \varepsilon_{LN}\, E_{LN} = \varepsilon_0\, \varepsilon_{SiO_2}\, E_{SiO_2} = \text{const}, \tag{2}$$

d_{LN} and d_{SiO_2} were the thicknesses of LN thin film and SiO$_2$ layer, which have been determined. Therefore, the electric field in the LN thin film layer is given by Equations (1) and (2).

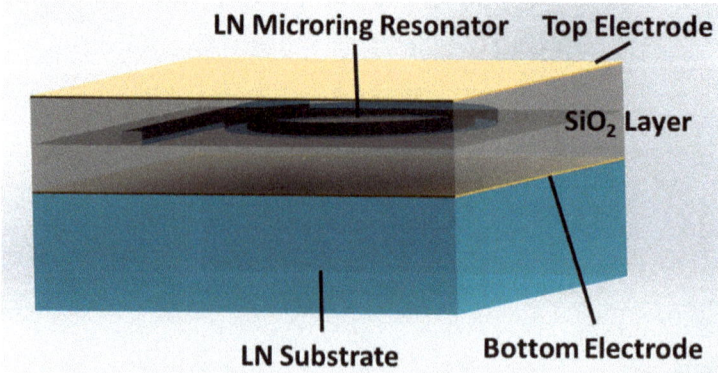

Figure 7. The LN microring resonator was embedded in the middle SiO$_2$ layer, and the electrodes were placed over below and above the SiO$_2$ layer.

Optical transmission simulations were performed to characterize the electrical tuning of optical resonances. To observe the electro-optic effect, an electric field applied between the top and the bottom electrodes by applying different direct current (DC) voltages was considered uniform. Figure 8 shows the simulated TM-mode spectrum as a function of different electric field intensities. The resonance wavelength shifted ascending with increasingly electric field intensity. The resonance shifts were 99 pm, 198 pm, 297 pm, and 395 pm for the chance in different electric field intensities from 0 V/μm to 1 V/μm, 2 V/μm, 3 V/μm, and 4 V/μm, respectively.

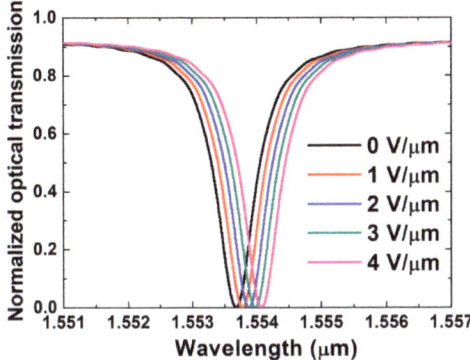

Figure 8. Transmission spectra of wavelength shift due to different electric field intensities in the Z-direction. The calculations referred to the TM mode at the microring radius R = 20 μm.

4. Conclusions

In conclusion, a VarFDTD method was used to design and simulate various waveguides and waveguide-coupled single-mode microring resonators in z-cut LNOI. Single mode conditions were obtained. The propagation losses of LN planar waveguides were calculated by full-vectorial finite difference method. The propagation losses of LN planar waveguides decreased with the increase of the SiO_2 cladding layer thickness. The thickness of LN Film, width of ridge waveguide and thickness of SiO_2 layer were optimized to 0.5 μm, 0.7 μm and 2 μm, respectively. Q-factor ascended with increasing radius and gap sizes, while FSR descended with increasing radius. A microring resonator with Q-factor more than 10,000 and FSR more than 8 nm was achieved in the case of gap = 0.29 μm and radius R = 20 μm. Our simulation showed that the resonance wavelength of the electro-optical microresonators in LNOI shifted ascending with increasingly electric field intensity. This work will continue with fabrication and characterization. For this purpose, the results provided here will be of useful guidance before laboratory works are carried out.

Author Contributions: B.X. conceived the original idea; H.H. carried out the simulations, analyzed the data, and wrote the paper; B.X. and J.Z. contributed the useful and deep discussions and modified the manuscript.

Funding: This work was supported by the National Natural Science Foundation of China (NSFC) (Grants NO. 61701288), the Shenzhen Science and Technology Planning (NO. JCYJ20170818143327496), the Foundation of Zibo Vocational Institute (NO. 2018zzzr03), and the Shandong University Science and Technology Planning (NO. J16LN93).

Conflicts of Interest: The authors declare no conflict of interest.

References

1. Weis, R.S.; Gaylord, T.K. Lithium niobate: Summary of physical properties and crystal structure. *Appl. Phys. A* **1985**, *37*, 191–203. [CrossRef]
2. Chen, L.; Wood, M.G.; Reano, R.M. 12.5 pm/V hybrid silicon and lithium niobate optical microring resonator with integrated electrodes. *Opt. Express* **2013**, *21*, 27003–27010. [CrossRef] [PubMed]

3. Rabiei, P.; Gunter, P. Optical and electro-optical properties of submicrometer lithium niobate slab waveguides prepared by crystal ion slicing and wafer bonding. *Appl. Phys. Lett.* **2004**, *85*, 4603–4605. [CrossRef]
4. Poberaj, G.; Hu, H.; Sohler, W.; Günter, P. Lithium niobate on insulator (LNOI) for micro-photonic devices. *Laser Photon. Rev.* **2012**, *6*, 488–503. [CrossRef]
5. Tao, S.H.; Mao, S.C.; Song, J.F.; Fang, Q.; Yu, M.B.; Lo, G.Q.; Kwong, D.L. Ultra-high order ring resonator system with sharp transmission peaks. *Opt. Express* **2010**, *18*, 393–400. [CrossRef] [PubMed]
6. Rafizadeh, D.; Zhang, J.P.; Hagness, S.C.; Taflove, A.; Stair, K.A.; Ho, S.T.; Tiberio, R.C. Waveguide-coupled AlGaAs/GaAs microcavity ring and disk resonators with high finesse and 21.6-nm free spectral range. *Opt. Lett.* **1997**, *22*, 1244–1246. [CrossRef] [PubMed]
7. Rao, A.; Fathpour, S. Compact lithium niobate electrooptic modulators. *IEEE J. Sel. Top. Quantum Elect.* **2018**, *24*, 1–14. [CrossRef]
8. Almeida, V.R.; Barrios, C.A.; Panepucci, R.R.; Lipson, M. All-optical control of light on a silicon chip. *Nature* **2004**, *431*, 1081–1084. [CrossRef] [PubMed]
9. Shahoei, H.; Dumais, P.; Yao, J. Continuously tunable photonic fractional Hilbert transformer using a high-contrast germanium-doped silica-on-silicon microring resonator. *Opt. Lett.* **2014**, *39*, 2778–2781. [CrossRef] [PubMed]
10. Rabiei, P.; Steier, W.H.; Zhang, C.; Dalton, L.R. Polymer micro-ring filters and modulators. *J. Lightwave Technol.* **2002**, *20*, 1968–1975. [CrossRef]
11. Guarino, A.; Poberaj, G.; Rezzonico, D.; Degl'innocenti, R.; Günter, P. Electro-optically tunable microring resonators in lithium niobate. *Nat. Photon.* **2007**, *1*, 407–410. [CrossRef]
12. Zhang, M.; Wang, C.; Cheng, R.; Shams-Ansari, A.; Lončar, M. Monolithic ultra-high-Q lithium niobate microring resonator. *Optica* **2017**, *4*, 1536–1537. [CrossRef]
13. Siew, S.Y.; Saha, S.S.; Tsang, M.; Danner, A.J. Rib microring resonators in lithium niobate on insulator. *IEEE Photon. Technol. Lett.* **2016**, *28*, 573–576. [CrossRef]
14. Chin, M.K.; Ho, S.T. Design and modeling of waveguide-coupled single-mode microring resonators. *J. Lightwave Technol.* **1998**, *16*, 1433–1446. [CrossRef]
15. Lalanne, P.; Hugonin, J.P. Bloch-wave engineering for high-Q, small-V microcavities. *IEEE J. Sel. Top. Quantum Elect.* **2003**, *39*, 1430–1438. [CrossRef]
16. Hazura, H.; Shaari, S.; Menon, P.S.; Majlis, B.Y.; Mardiana, B.; Hanim, A.R. Design parameters investigation of single mode silicon-on-insulator (SOI) microring channel dropping filter. *Adv. Sci. Lett.* **2013**, *19*, 199–202. [CrossRef]
17. Hu, H.; Yang, J.; Gui, L.; Sohler, W. Lithium niobate-on-insulator (LNOI): Status and perspectives. *Proc. SPIE.* **2012**, *8431*, 84311D.
18. Rao, A.; Fathpour, S. Heterogeneous thin-film lithium niobate integrated photonics for electrooptics and nonlinear optics. *IEEE J. Sel. Top. Quantum Elect.* **2018**, *24*, 8200912. [CrossRef]
19. Chen, Z.; Wang, Y.; Jiang, Y.; Kong, R.; Hu, H. Grating coupler on single-crystal lithium niobate thin film. *Opt. Mater.* **2017**, *72*, 136–139. [CrossRef]
20. Lumerical Solutions. Available online: http://www.lumerical.com/ (accessed on 4 August 2018).
21. Schlarb, U.; Betzler, K. A generalized sellmeier equation for the refractive indices of lithium niobate. *Ferroelectrics* **1994**, *156*, 99–104. [CrossRef]
22. Chen, Z.; Peng, R.; Wang, Y.; Zhu, H.; Hu, H. Grating coupler on lithium niobate thin film waveguide with a metal bottom reflector. *Opt. Mater. Express* **2017**, *7*, 4010–4017. [CrossRef]
23. Han, H.; Cai, L.; Xiang, B.; Jiang, Y.; Hu, H. Lithium-rich vapor transport equilibration in single-crystal lithium niobate thin film at low temperature. *Opt. Mater. Express* **2015**, *5*, 2634–2641. [CrossRef]

© 2018 by the authors. Licensee MDPI, Basel, Switzerland. This article is an open access article distributed under the terms and conditions of the Creative Commons Attribution (CC BY) license (http://creativecommons.org/licenses/by/4.0/).

Article

Investigation of HfO$_2$ Thin Films on Si by X-ray Photoelectron Spectroscopy, Rutherford Backscattering, Grazing Incidence X-ray Diffraction and Variable Angle Spectroscopic Ellipsometry

Xuguang Luo [1], Yao Li [1], Hong Yang [1], Yuanlan Liang [1], Kaiyan He [1,*], Wenhong Sun [2,*], Hao-Hsiung Lin [3], Shude Yao [4], Xiang Lu [1], Lingyu Wan [1,*] and Zhechuan Feng [1,*]

[1] Laboratory of Optoelectronic Materials & Detection Technology, Guangxi Key Laboratory for the Relativistic Astrophysics, School of Physical Science & Technology, Guangxi University, Nanning 530004, China; lxg142234@139.com (X.L.); liyao13@mail.gxu.cn (Y.L.); 18672513177@163.com (H.Y.); LYL199245@163.com (Y.L.); luxiang@gxu.edu.cn (X.L.)
[2] School of Physical Science & Technology, Guangxi University, Nanning 530004, China
[3] Department of Electrical Engineering, Graduate Institute of Electronics Engineering, National Taiwan University, Taipei 106-17, Taiwan; hhlin@ntu.edu.tw
[4] State Key Laboratory of Nuclear Physics and Technology, Peking University, Beijing 100871, China; sdyao@pku.edu.cn
* Correspondence: gredhky@gxu.edu.cn (K.H.); youzi7002@gxu.edu.cn (W.S.); wanlingyu75@126.com (L.W.); fengzc@gxu.edu.cn (Z.F.)

Received: 23 April 2018; Accepted: 5 June 2018; Published: 12 June 2018

Abstract: Hafnium oxide (HfO$_2$) thin films have been made by atomic vapor deposition (AVD) onto Si substrates under different growth temperature and oxygen flow. The effect of different growth conditions on the structure and optical characteristics of deposited HfO$_2$ film has been studied using X-ray photoelectron spectroscopy (XPS), Rutherford backscattering spectrometry (RBS), grazing incidence X-ray diffraction (GIXRD) and variable angle spectroscopic ellipsometry (VASE). The XPS measurements and analyses revealed the insufficient chemical reaction at the lower oxygen flow rate and the film quality improved at higher oxygen flow rate. Via GIXRD, it was found that the HfO$_2$ films on Si were amorphous in nature, as deposited at lower deposition temperature, while being polycrystalline at higher deposition temperature. The structural phase changes from interface to surface were demonstrated. The values of optical constants and bandgaps and their variations with the growth conditions were determined accurately from VASE and XPS. All analyses indicate that appropriate substrate temperature and oxygen flow are essential to achieve high quality of the AVD-grown HfO$_2$ films.

Keywords: hafnium dioxide (HfO$_2$); X-ray photoelectron spectroscopy (XPS); Rutherford backscattering spectrometry (RBS); grazing incidence X-ray diffraction (GIXRD); variable angle spectroscopic ellipsometry (VASE)

1. Introduction

A variety of transistors is beneficial to enhance the function and performance of the integrated circuit (IC). The IC transistor's channel length and gate dielectric thickness and other device dimensions shrink rapidly. However, the conventional silicon dioxide (SiO$_2$) gate dielectric is a constraint because the direct tunnel leakage current and the rate of dissipation increase with the decrease of layer thickness [1–5]. Accordingly, new materials with the high dielectric constant (high-k materials) replacing SiO$_2$ as gate dielectrics are developed, in order to improve the device performance and reduce the leakage currents [2–5].

Many high-k materials have been used as gate oxide, such as HfO_2, ZrO_2, Y_2O_3, La_2O_3, Si_3N_4, TiO_2, and Al_2O_3. Hafnium dioxide (HfO_2) possesses promising properties of high dielectric constant (k ~25), relatively wide bandgap of around 5.8 eV, a good thermal stability [5], a high breakdown electric field (5 MV/cm), and good thermal stability on Si substrate [2]. Therefore, HfO_2 has been the dielectric material in processor transistors for more than ten years already, and is an excellent material for metal oxide semiconductor (MOS) based microelectronic devices [3], resistive switching material in memory devices [4], and in optical coatings [2]. As such, it has been studied extensively both experimentally and theoretically [1–24]. However, in electronic applications, the frequent coercion of stoichiometric deviations exists on thin films of HfO_2 which are typically sputtered or grown by atomic layer deposition (ALD) [2,6]. Atomic vapor deposition (AVD), as a special MOCVD process mode, is used to deposit pure HfO_2 at a wide range of temperatures [7], and enables high gas-phase saturation of the precursors, high growth rate, and full stoichiometric control of films [8]. We mainly analyze the effect of different growth conditions on the structure, and the stoichiometric and optical characteristics by using AVD for HfO_2 thin films.

In this paper, a series of nanometer scale (33–70 nm) HfO_2 thin films grown on Si substrates under different conditions by AVD are studied. With the help of X-ray photoelectron spectroscopy (XPS), Rutherford backscattering spectrometry (RBS), grazing incidence X-ray diffraction (GIXRD), and variable angle spectroscopic ellipsometry (VASE), the composition, crystallization phases, and optical constants of the films are characterized and penetratively studied from surface to interface.

2. Materials and Methods

Firstly, the p-type (100) silicon wafers were cleaned by HF-dipping. Then, HfO_2 films with different thicknesses of about 30–70 nm were deposited by atomic vapor deposition (AVD) using an AIXTRON Tricent system with oxygen flow of 500 sccm and 800 sccm at a substrate temperature of 400 °C and 500 °C. Hafnium diethylamide, $Hf[N(C_2H_5)_2]_4$, was the source material for the AVD of HfO_2 [9]. Argon was used as transmission gas with flow rate of 200 sccm, and the total pressure in the chamber was fixed at 5 mbar and the injection frequency is 3 Hz, as in another previous growth procedure [10]. Further details for the growth of HfO_2 can be found there [9,10]. In this study, three kinds of HfO_2 films were prepared on 2″ Si wafers by AVD, with thicknesses of approximately 70, 33, and 34 nm, respectively, which were measured via transmission electron microscopy (TEM).

Three samples of S1, S2 and S3 with the original run numbers of Hf08a, Hf08b and Hf08c, respectively, their information, and growth conditions, are listed at Table 1.

Table 1. Growth conditions and measured thicknesses of HfO_2 thin films by TEM, from the grower.

Sample	Structure	Growth Conditions	Thicknesses
S1 (Hf08a)	HfO_2/Si	400 °C, 5 mbar, O_2: 500 sccm	70 nm
S2 (Hf08b)	HfO_2/Si	500 °C, 5 mbar, O_2: 500 sccm	33 nm
S3 (Hf08c)	HfO_2/Si	500 °C, 5 mbar, O_2: 800 sccm	34 nm

XPS (also known as electron spectroscopy for chemical analysis—ESCA) uses highly focused monochromatic X-ray to probe the material of interest. The energy of the photoemitted electrons ejected by the X-rays is specific to the chemical state of the elements and compounds presented, i.e., bound state or multivalent state of individual elements can be differentiated. In this work, the chemical states in the surface region of HfO_2 thin films deposited on Si substrate were studied by XPS, with the incident Al $K\alpha$ beam at the energy of 1486.6 eV. For the thin films analyzed, all the energy scales of the XPS spectra were calibrated by the binding energy (B.E.) of the C 1s peak at 285.0 eV [11]. Afterward, the selected peaks are fitted by the fitting program XPSPEAK4.1. The energy bandgap values of thin oxides can be determined from the photoelectron loss-energy spectra calculated from the XPS O 1s peaks [25,26].

Rutherford backscattering spectrometry (RBS) was employed to determine the contents of films nondestructively. The random spectra of films are recorded via RBS, and simulated with the software of simulation of ions in matter and nuclear reaction analysis (SIMNRA) [12]. The crystal structure phases of films were scanned by grazing incidence X-ray diffraction (GIXRD) with Cu Kα radiation (0.15418 nm). The diffraction angle, 2θ, was varied from 20° to 52°, and incident angles were set as 0.5°, 1°, 3°, and 5°, respectively. The crystallite size of HfO_2 thin film of S2 and S3 with different incident angles has been calculated.

The variable angle spectroscopic ellipsometry (VASE) measurements were performed by a dual rotating compensator Mueller matrix ellipsometer (ME-L ellipsometer, Wuhan Eoptics Technology Co. Ltd., Wuhan, China) in the spectral range from 195 to 1680 nm (0.74–6.35 eV) with 1 nm step interval at five angles of incidence (45°, 50°, 55°, 60°, 65°, 70°). We adopt the Tauc–Lorentz dispersion function to extracted dielectric functions [13–16]. The analyses of VASE experiment data helped us to fit out the optical constants, the optical bandgaps, as well as information on thickness and roughness of HfO_2 thin films.

3. Results and Discussion

3.1. X-ray Photoelectron Spectroscopy (XPS)

3.1.1. XPS Survey Scan

XPS survey scans were performed for all three HfO_2 samples. Figure 1 shows such a typical wide survey scan for a sample S3 (Hf08c). The characteristic peaks of Hf (4f, 4d, 4p), O 1s, C 1s, N 1s, and O KLL (Auger peaks) are observed in the general survey spectra. None of contamination species have been observed within the sensitivity of the instrument expecting the adsorbed atmospheric carbon and nitrogen on the surface [11].

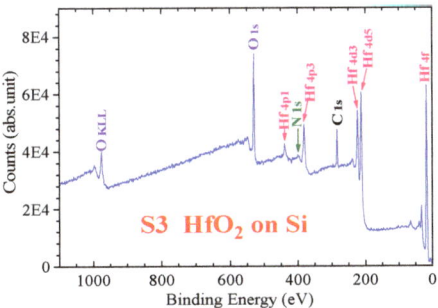

Figure 1. XPS survey scan of the HfO_2 film of the sample S3.

3.1.2. XPS Fitting and the Elements Composition Calculating

High-resolution XPS scans on the Hf 4f and O 1s peaks were performed for three HfO_2 films, as shown in Figure 2a–c. We focus on the analyses of the asymmetric shape of the Hf 4f and O 1s spectra to reveal the relative atomic percentage and bonding phase in the surface region. The background subtraction was carried out using Shirley's iterative method. The baseline of background is a combination of Shirley and linear function. The fitting program XPSPEAK4.1 was used to fit the experimental curves with a Gaussian–Lorenzian mixed function [17]. From the Hf 4f doublet, Hf $4f_{7/2}$ and Hf $4f_{5/2}$, the spin-orbit splitting and the intensity ratio of the components were set at 1.67 eV and 4/3, respectively [18].

Details of the Hf 4f fine spectra of three samples are shown in Figure 2a–c, respectively. The Hf 4f spectrum could be fitted with two sets of double-peak components. One set at 17.16 eV and 18.83 eV correspond to the Hf^{4+} 4f$_{7/2}$ and Hf^{4+} 4f$_{5/2}$ peaks of the Hf oxide bond (O–Hf–O), respectively [11]. While another set at 16.50 eV and 18.26 eV correspond to the Hf^{x+} 4f$_{7/2}$ and Hf^{x+} 4f$_{5/2}$ peaks (x < 4) of the Hf suboxide bond, respectively [19,20]. It is obvious that the doublet peaks of the suboxidized Hf^{x+} are stronger than those of the fully oxidized Hf^{4+} from the Figure 2a,b. However, the situation is opposite in Figure 2c. The full width at half maximum (FWHM) of fitted peaks are shown in Table 2. In comparison, the corresponding FWHMs of the Hf^{4+} in Figure 2a,b are wider than that in Figure 2c. This increased width was caused from the presence of a higher degree of disorder in the deficient films or from the coexistence of tetragonal and monoclinic phases with small differences in binding energy [18]. Our GIXRD data also support this scenario, with discussion in Section 3.3.

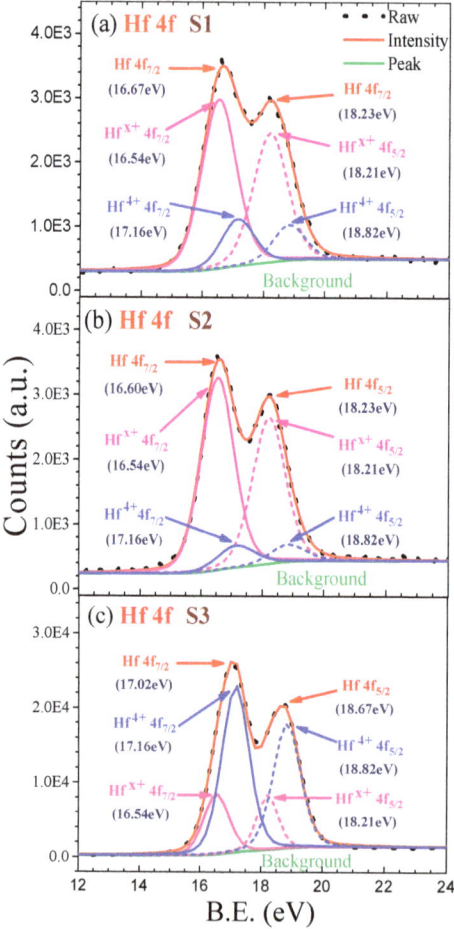

Figure 2. The high-resolution XPS fitting results of the Hf 4f peaks of S1, S2, and S3 are shown in (**a**–**c**), respectively, with experiment data (short dash lines), fitting result (red solid lines) and background (chartreuse solid lines).

A number of factors could lead to the binding energy shifts, such as charge transfer effect, environmental charge density, presence of electric field, and hybridization. Among these factors, charge transfer causing a binding energy shift is regarded as the dominant mechanism for the S2 and S3, with different oxygen flow rates in growth conditions [21]. According to the charge transfer mechanism, an electron removed from the valence orbital generates the increment in core electron's potential, and finally leads to a chemical binding energy shift [22]. The difference between the sample S2 in Figure 2b and the S3 in Figure 2c is attributed to the difference of the oxygen flow rates. In the deposition process of the S3, Hf oxide bonds were dominate over Hf suboxide bonds, due to the excess oxygen at 800 sccm oxygen flow rate. On the contrary, the sample S2 with 500 sccm oxygen flow rate exhibited a reverse trend: Hf suboxide bonds dominating over Hf oxide bonds. Therefore, it is reasonable that the peaks shift, from 16.60 to 17.02 eV for Hf $4f_{7/2}$, and from 18.23 to 18.67 eV for Hf $4f_{5/2}$, as shown in Figure 2b,c, can be explained from the enhanced charge transfer with the increase of the O_2 gas flow [23]. The intensity change between the Hf^{4+} 4f and Hf^{x+} 4f peaks can be demonstrated from the chemical shift in the binding energy of the Hf 4f peaks with oxygen flow rate. Both the Hf^{x+} $4f_{7/2}$ and Hf^{x+} $4f_{5/2}$ peaks in Figure 2a,b have similarly dominated components for the Hf $4f_{7/2}$ and Hf $4f_{5/2}$ peaks, respectively, while they are weaker than the Hf^{4+} $4f_{7/2}$ and Hf^{4+} $4f_{5/2}$ peaks, respectively in Figure 2c.

In Figure 3a–c, the XPS fitting results on the O 1s peaks of S1, S2, and S3 are shown, respectively. The O 1s peak actually consist of two very closely peaks, the Hf–O bond of O–Hf–O at 530.28~530.40 eV, and the O–O bond of non-lattice oxygen at 532.08 eV, which was contributed to by the suboxides with Hf [17,24,27,28]. The Hf–O bond belongs to the lattice oxygen which dominates in the Figure 2c. This demonstrates that the corresponding samples have ordered structures and good qualities at surface. In the Figure 2a,b, the high percentage of the non-lattice oxygen peak at 532.08 eV from the O 1s spectra indicates more defects existed in the HfO_2 film layer surface of S1 and S2.

Comparing the areas of O 1s and Hf 4f peaks, the stoichiometric composition ratio of O and Hf elements at the surface layer of the HfO_2 film can be estimated from the XPS fitting results. The atomic percentages and the FWHMs of the XPS O 1s and Hf 4f peaks for HfO_2 films deposited on Si are summarized in Table 2. There are some differences among these three samples in their XPS fitting results. The compositions of Hf are 31.63%, 33.10%, and 36.95% for S1, S2 and S3, respectively. The surface ratios of O and Hf elements of the three samples are between 1.70 and 2.13, which are in accordance with the range calculated by Myoung-Seok Kim et al. [23]. The higher proportions of oxygen in the sample of S1 and S2 are due to the presence of the non-lattice oxygen.

Table 2. The Hf and O atomic percentages and bonding analyses of S1, S2 and S3.

Sample		Hf^{x+} $4f_{7/2}$	Hf^{x+} $4f_{5/2}$	Hf^{4+} $4f_{7/2}$	Hf^{4+} $4f_{5/2}$		O	O–Hf
S1	Peak/eV	16.54	18.21	17.16	18.82	Peak/eV	532.08	530.27
	FWHM	1.34	1.32	1.27	1.25	FWHM	1.93	1.65
	Area/ASF (2.05)	4118.46	3088.84	1083.73	812.80	Area/ASF (0.66)	2400.19	3935.01
	Atom Hf (%)	0.14	0.11	0.04	0.03	Atom O (%)	0.26	0.42
			31.63%				68.37%	
S2	Peak/eV	16.54	18.21	17.16	18.82	Peak/eV	532.08	530.28
	FWHM	1.28	1.27	1.45	1.39	FWHM	1.96	1.57
	Area/ASF (2.05)	4425.81	3319.36	565.70	424.27	Area/ASF (0.66)	1822.13	3885.52
	Atom Hf (%)	0.17	0.13	0.02	0.02	Atom O (%)	0.21	0.46
			33.01%				66.99%	
S3	Peak/eV	16.54	18.21	17.16	18.82	Peak/eV	532.08	530.33
	FWHM	0.97	0.80	1.06	1.06	FWHM	1.38	1.67
	Area/ASF (2.05)	9172.99	6879.74	27,466.56	20,599.92	Area/ASF (0.66)	2901.01	32,328.96
	Atom Hf (%)	0.05	0.04	0.16	0.12	Atom O (%)	0.05	0.58
			36.95%				63.05%	

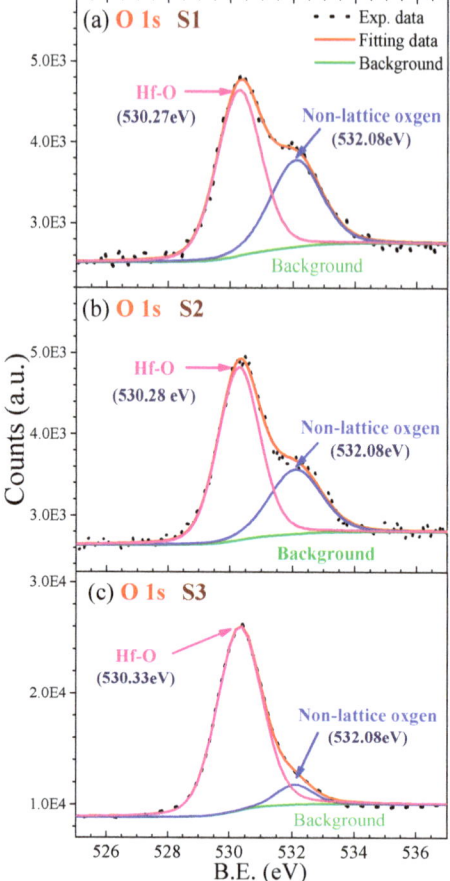

Figure 3. The high-resolution XPS fitting results of the O 1s peaks of S1, S2, and S3 are shown in (**a–c**), respectively, with experiment data (short dash lines), fitting result (red solid lines), and background (chartreuse solid lines).

3.1.3. Energy Bandgaps of HfO$_2$ Deduced from XPS

The energy bandgap values of thin oxides can be deduced from the energy loss signals for O 1s photoelectrons [25,26]. The photoelectron loss-energy spectra of three HfO$_2$ films are shown in Figure 4. The start of the photoelectron loss-energy spectrum, after setting the energy of the O 1s peak maximum to zero loss energy, was defined by linearly extrapolating the segment of maximum negative slope (dash-lines in the Figure 4) to the background level in each of the spectra [29–31]. The bandgap values of S1, S2, and S3 are as follows: 5.65 eV, 5.57 eV and 5.10 eV, which are in agreement with our results by variable angle spectroscopic ellipsometry (VASE) in Section 3.4, and previously reported bandgap values for HfO$_2$ in different references [3,32–37].

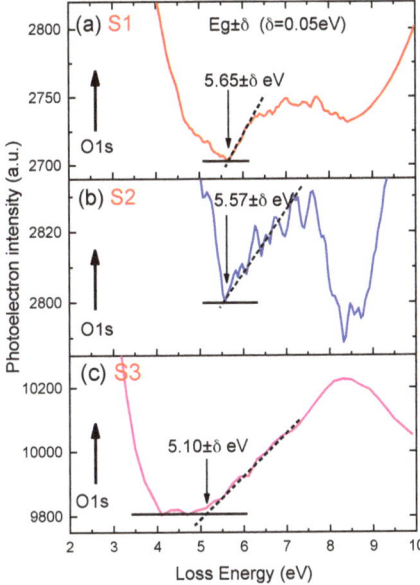

Figure 4. The photoelectron energy-loss spectra from the O 1s peaks of HfO$_2$ films for S1, S2 and S3 by XPS are shown in (**a**–**c**), respectively.

3.2. Rutherford Backscattering Spectrometry (RBS)

The random RBS spectra for three HfO$_2$ layers on Si and their corresponding simulated fits by SIMNRA are given in Figure 5. The contents of three HfO$_2$ films are determined precisely and listed in Table 3, and the calculated film thicknesses are shown in Table 5. The arrows (labeled with Hf, O, and Si) in Figure 5 denote the energy for backscattering from Hf, O, and Si atoms, respectively. The perfectly symmetrical shapes of the hafnium peaks indicate a negligible film roughness of the three samples with respect to the mean thickness [38].

Although the RBS probes the entire film and the XPS mainly the surface area of only about 5–10 nm, they are particularly close to the expected ideal composition value of Hf$_{0.33}$O$_{0.67}$. The measurements of XPS and RBS suggest that the stoichiometric HfO$_2$ films were deposited dominantly on Si. However, there is still a little deviation between the simulated compositions by RBS and XPS for all samples. For the S1 and S2 films, from both RBS and XPS, the ratios (Hf:O) are 1:N (N > 2). There exists a bit over stoichiometry oxygen, which might be caused from the non-lattice oxygen, as discussed in Section 3.1.2. For the film S3, the ratios (Hf:O) from both RBS and XPS are 1:N (N < 2), indicating a small over stoichiometry hafnium, which might be attributed from the sub-oxidized Hf^{x+} (x < 4), such as the theoretically predicted semi-metallic Hf$_2$O$_3$ [9,39].

Table 3. The compositions of the three HfO$_2$ films on Si obtained by RBS and XPS.

Samples	Hf (Composition)		O (Composition)		Ratios (Hf:O)	
	RBS	XPS	RBS	XPS	RBS	XPS
S1 (Hf08a)	0.30	0.32	0.70	0.68	1:2.33	1:2.13
S2 (Hf08b)	0.34	0.33	0.66	0.67	1:1.94	1:2.03
S3 (Hf08c)	0.35	0.37	0.65	0.63	1:1.86	1:1.70

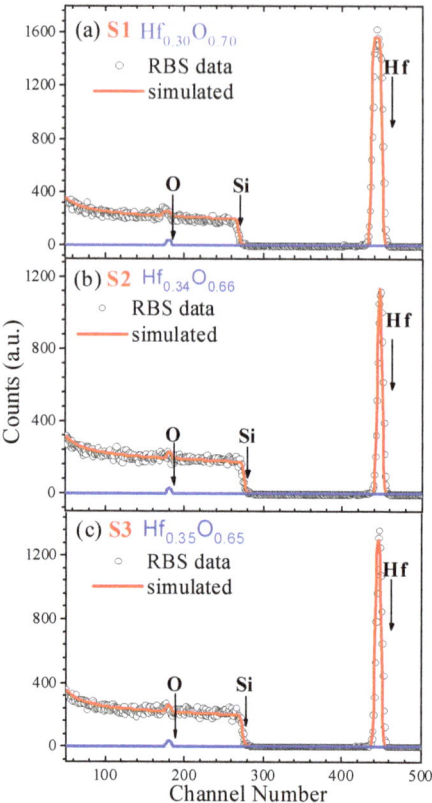

Figure 5. Random (open circles) and simulated fitting (solid lines) RBS spectra of of S1, S2 and S3 are shown in (**a**–**c**), respectively.

3.3. Grazing Incidence X-ray Diffraction (GIXRD)

Grazing incidence X-ray diffraction (GIXRD) patterns of S1, S2, and S3 with different incident angles (0.5°, 1°, 3°, and 5°) are shown in Figure 6, respectively. These spectra demonstrate the crystallization and the crystallization phases of the thin HfO_2 films, and variations from the surface to the interface corresponding to the incident angle from 0.5° to 5°. The broad GIXRD curves in Figure 6a indicate that the HfO_2 film of S1 is amorphous in nature. This is due to the fact that the low deposition temperature of 400 °C cannot provide sufficient energy to form a crystalline HfO_2 layer by AVD [40–42]. The GIXRD patterns of samples S2 and S3 deposited at substrate temperature of 500 °C indicate that these HfO_2 films are polycrystalline. The peaks of GIXRD patterns in Figure 6b,c are indexed by the monoclinic, tetragonal, and orthorhombic phases, correspondingly. It is noted that orientation indexes of (−111), (002), (200), (−211), (211), and (−221) correspond to the dominant monoclinic phase of HfO_2 film. The low intensity peak at 2θ = 30.5° refers to the (−111) orthorhombic phases [43,44]. There are two peaks at 31.0° and 32.3°, being attributed to the (002) tetragonal and (011) tetragonal phases of Hf_2O_3, respectively, according to the calculated data by Kan-Hao Xue et al. [39].

In the Figure 6b,c, the peaks of monoclinic, tetragonal, and orthorhombic phases are clearly seen at 3° and 5° incidences. When the detection angles were at 1° and 0.5°, the m(−111) peaks of monoclinic phase are dominant, and the peaks of orthorhombic and tetragonal phase are weak to hardly visible. In other words, there are three phases deposited at the initial stage of film growth. As the thickness increases, the tetragonal and orthorhombic phases gradually weaken to almost invisible. Recently, it was reported that HfO$_2$ thin films could be crystallized in both orthorhombic and tetragonal phases when they are sufficiently thin (<10 nm) and the grain size is small [45,46]. Our investigation revealed that the HfO$_2$ layer had all monoclinic, tetragonal, and orthorhombic structural phases at its initial growth stage, but as they grow thicker, the tetragonal and orthorhombic phases become weaker, and finally almost disappeared.

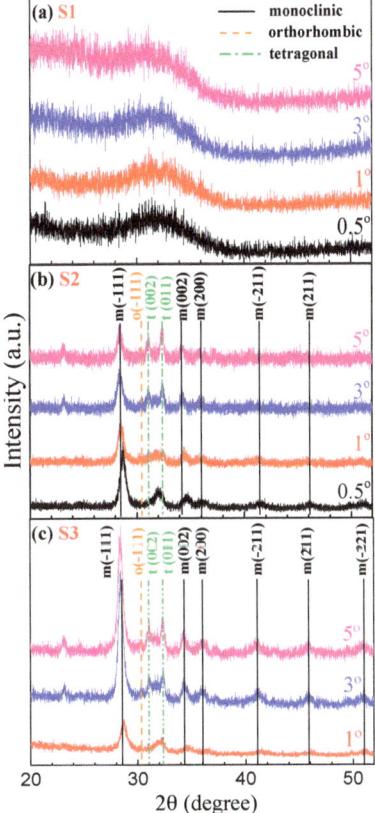

Figure 6. Grazing incidence X-ray diffraction (GIXRD) patterns of S1, S2 and S3 with different incident angles (0.5°, 1°, 3° and 5°) are shown in (**a**–**c**), respectively.

By the Scherrer formula, the crystallite size of HfO$_2$ thin film of S2 and S3 with different incident angles can be calculated using the m(−111) peak of the monoclinic phase [47–49].

$$D = \frac{k \cdot \lambda}{\beta \cos \theta}, \tag{1}$$

where D is the crystallite size, λ is the X-ray wavelength of Cu $K\alpha$ (0.15418 nm), k is the Scherrer constant of the order of unity (0.95 for powder and 0.89 for film), β is the full width of peak at half maximum intensity (FWHM), and θ is the corresponding Bragg diffraction angle [47–49].

Figure 7 exhibits the m(−111) peak patterns of monoclinic phase of S2 and S3 with incident angles of 0.5°, 1°, 3° and 5°, from Figure 6b,c, respectively. The values of the peak position, FWHM and the calculated crystallite size are listed in Table 4. As the incident angle varies from 5° to 0.5°, i.e., decreasing the depth of detection, the m(−111) peak positions were shifted towards the high angle side and their FWHMs were increased, both gradually. Also, the crystallite size became smaller with decreasing the depth of detection. Generally, the crystal lattice with dwindling crystal size tends to generate phases of higher symmetry [50,51]. These correspond to the appearance of tetragonal and orthorhombic phases in Figure 6b,c. Multiple mixed crystal phases result in the lower symmetry of the HfO_2 in the depths.

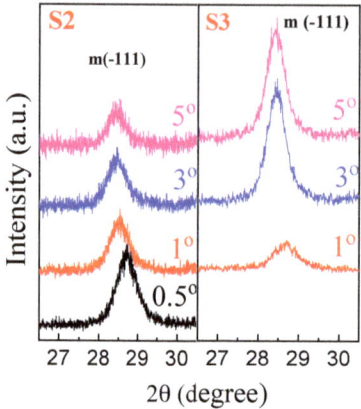

Figure 7. The m(−111) peak of monoclinic phase of S2 and S3 in GIXRD patterns.

Table 4. Peak position (2θ (degree)), FWHM (β (degree)), and crystallite size (D (nm)). The average errors estimations are 0.004°, 0.007°, and 0.008 nm, respectively.

Incident Angles	S2			S3		
(Degree)	2θ (Degree)	β (Degree)	D (nm)	2θ (Degree)	β (Degree)	D (nm)
0.5°	28.720	0.548	14.810			
1°	28.521	0.521	15.570	28.680	0.605	13.413
3°	28.460	0.513	15.811	28.444	0.479	16.932
5°	28.449	0.460	17.632	28.426	0.472	17.183

In Table 4, the relatively similar data on the crystallite size (D) for S2 and S3 show that there is no obvious difference in microcosmic. However, the sharp and intense peaks of S3 in the XRD spectra, shown in Figures 6 and 7, demonstrate its structure is more orderly than S2 on a large scale. Among the GIXRD patterns of S1, S2 and S3, the structure of S3 is the most orderly, and its quality is the best among these three samples, which are consistent with our results from XPS and RBS measurements and analyses.

3.4. The Variable Angle Spectroscopic Ellipsometry (VASE)

The best fit results of VASE spectra for HfO_2/Si sample S3 are presented in Figure 8, where dot lines depict SE experiment data, and solid lines are the fitting results. The data displayed for two key ellipsometry parameters, Psi (Ψ) and Delta (Δ), are related to the change in amplitude and phase shift of the impinging E field upon reflection, and are wavelength dependent [52]. In our analyses, we have considered an interface region with silicate (SiO_x), which is observed clearly in the interface by high-resolution transmission electron microscope (HR-TEM), as shown in Figure 9. The surface roughness was described by Bruggeman effective medium approximation (EMA) [52,53]. The fitting structural model were constructed as silicon substrate/interface layer/HfO_2 layer/surface roughness. In order to get more reliable fitting results, the VASE spectra were measured with five incident angles of 50°, 55°, 60°, 65°, and 70° for the three samples. The fitted thicknesses of roughness, HfO$_2$ layer, and interface layer for three samples were obtained from VASE.

In Table 5, the thicknesses of the corresponding layer of S1, S2 and S3 are close from different measurements, including VASE, RBS, HR-TEM, and the measured thicknesses from the grower. It is more accurate about the thicknesses from HR-TEM and VASE. In comparison with the interface thicknesses of three samples, higher substrate temperatures, and larger oxygen flow in growth progress result in thicker interface of silicate.

The fitted optical constants of refractive index n and extinction coefficient k are presented in Figure 10. As the photon energy is less than 4.70 eV, the HfO$_2$ films are at the transparent region. However, the extinction coefficients start to increase rapidly at the high-energy edge from 4.70 eV, resulting from the remarkable subgap absorption [16].

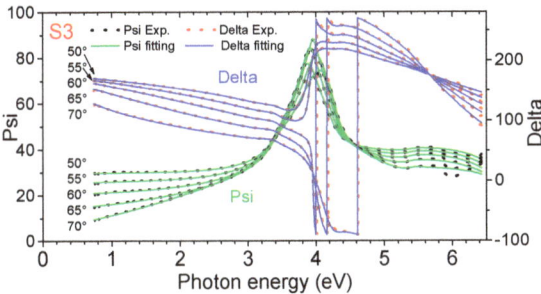

Figure 8. SE spectra (dot lines) and fitted (solid lines) psi and delta spectra vs photon energy with five incident angles (50°, 55°, 60°, 65° and 70°) of HfO$_2$/Si sample S3 at RT (300 K).

Table 5. The thicknesses of three samples are obtained by VASE, RBS, HR-TEM, and the measured thicknesses from grower.

Samples	Thickness (nm)						
	SE			RBS	HR-TEM		Grower
	Roughness	HfO$_2$	Interface	HfO$_2$	HfO$_2$	Interface	HfO$_2$
S1	4.71 ± 0.06	65.74 ± 0.08	1.90 ± 0.08	53.33	62.5	2.0	70
S2	3.95 ± 0.05	32.14 ± 0.11	2.59 ± 0.11	32.67			33
S3	4.01 ± 0.05	31.71 ± 0.10	3.37 ± 0.10	40.67			34

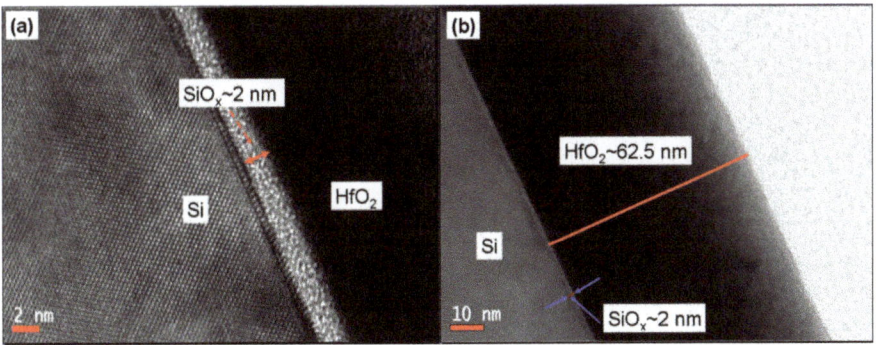

Figure 9. HR-TEM images of the S3 with scale of (**a**) 2 nm, and (**b**) 10 nm, respectively.

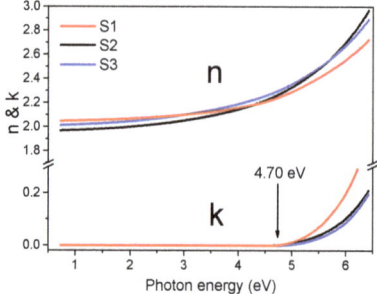

Figure 10. Optical constants of refractive indices (n) and extinction coefficients (k) for three HfO$_2$ films of S1, S2, and S3 at RT (300 K).

The HfO$_2$ films are known to have an indirect bandgap. Its optical bandgap versus photon energy hv is proportional to $(\alpha hv)^{1/2}$, ($(\alpha hv)^2$ for direct bandgap material), where α and hv are the absorption coefficient and photon energy, respectively [3,14,54–56]. Figure 11 shows plots of $(\alpha hv)^{1/2}$ versus photon energy for our three HfO$_2$ films, respectively. The values of the optical bandgap E_g were calculated by the liners of $(\alpha hv)^{1/2}$ vs. hv extrapolated to the intersection with the photon energy axis. The bandgaps of HfO$_2$ films of S1, S2 and S3 were determined to be 5.35 eV, 5.34 eV, and 5.26 eV, respectively. These data are approximate to the values of bandgap obtained from the photoelectron energy-loss spectra by XPS in Section 3.1. Associated with the analysis results of GIXRD, the better quality of samples are related to the smaller bandgaps.

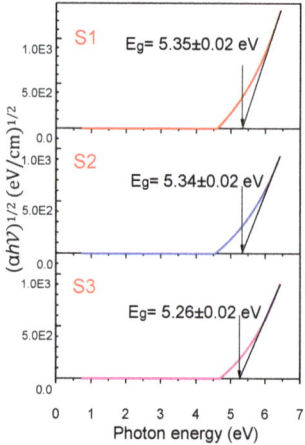

Figure 11. $(\alpha hv)^{1/2}$ versus photon energy plots of the HfO$_2$ films.

4. Conclusions

Comprehensive analyses have been conducted on a series of HfO$_2$ thin (30–70 nm) films grown on Si by AVD under different growth conditions via XPS, RBS, GIXRD, and VASE. The film characteristic parameters of thickness, structures, optical constants, bandgaps etc. were accurately determined. Important and significant results are obtained:

i. Through XPS measurements and analyses on the Hf 4f peaks and the change of intensity ratio between Hf^{4+} and Hf^{x+} peaks, it was revealed that at the lower oxygen flow rate of 500 sccm,

ii. The deposited HfO$_2$ films on Si were amorphous in nature at the low deposition temperature of 400 °C by AVD, while at higher deposition temperature, polycrystalline HfO$_2$ films were achieved.
iii. At the initial stage of film growth, the monoclinic, tetragonal, and orthorhombic phases co-existed. As the film grew thicker, the tetragonal and orthorhombic phases gradually weakened until the monoclinic phase dominated. The crystallite size of HfO$_2$ film became smaller from interface to surface, confirmed using varied angle GIXRD.
iv. It was found that for HfO$_2$ film, higher crystallization and more ordered structure correspond to a smaller bandgap, determined from VASE and XPS, close to single crystal HfO$_2$.

The comprehensive studies demonstrate that appropriate substrate temperature and oxygen flow are essential to the structure, chemical composition, and optical constants from surface and interface of the HfO$_2$ films deposited by AVD. This work with integrated experiment measurements and analyses has enhanced our understanding of AVD-grown HfO$_2$ advanced materials.

Author Contributions: Conceptualization, X.L., X.L., L.W. and Z.C.F.; Data curation, X.L.; Formal analysis, X.L., Y.L. and H.Y.; Funding acquisition, L.W. and W.S.; Investigation, X.L. and Y.L.; Project administration, Z.C.F.; Resources, H.-H.L., S.Y. and Z.C.F.; Supervision, Z.C.F.; Writing—original draft, X.L.; Writing—review & editing, X.L., Y.L., H.Y., L.W., W.S., K.H. and Z.C.F.

Acknowledgments: This work was supported by the National Natural Science Foundation of China [Nos. 61367004 and 61504030]; and the special funding [Nos. T3120097921 and T3120099202] for Guangxi distinguished [Bagui Rencai & Bagui Xuezhe].

Conflicts of Interest: The authors declare no conflict of interest.

References

1. Lu, Q.; Huang, R.; Lan, X.; Chi, X.; Lu, C.; Li, C.; Wu, Z.; Li, J.; Han, G.; Yan, P. Amazing diffusion depth of ultra-thin hafnium oxide film grown on n-type silicon by lower temperature atomic layer deposition. *Mater. Lett.* **2016**, *169*, 164–167. [CrossRef]
2. Gao, L.; Yalon, E.; Chew, A.R.; Deshmukh, S.; Salleo, A.; Pop, E.; Demkov, A.A. Effect of oxygen vacancies and strain on the phonon spectrum of HfO$_2$ thin films. *J. Appl. Phys.* **2017**, *121*, 224101. [CrossRef]
3. Fan, X.; Liu, H.; Zhong, B.; Fei, C.; Wang, X.; Wang, Q. Optical characteristics of H$_2$O-based and O$_3$-based HfO$_2$ films deposited by ALD using spectroscopy ellipsometry. *Appl. Phys. A* **2015**, *119*, 957–963. [CrossRef]
4. Shandalov, M.; McIntyre, P.C. Size-dependent polymorphism in HfO$_2$ nanotubes and nanoscale thin films. *J. Appl. Phys.* **2009**, *106*, 084322. [CrossRef]
5. Kim, J.C.; Heo, J.S.; Cho, Y.S.; Moon, S.H. Atomic layer deposition of an HfO$_2$ thin film using Hf(O-iPr)$_4$. *Thin Solid Films* **2009**, *517*, 5695–5699. [CrossRef]
6. Monaghan, S.; Hurley, P.K.; Cherkaoui, K.; Negara, M.A.; Schenk, A. Determination of electron effective mass and electron affinity in HfO$_2$ using MOS and MOSFET structures. *Solid-State Electron.* **2009**, *53*, 438–444. [CrossRef]
7. Cosnier, V.; Dabertrand, K.; Blonkowski, S.; Lhostis, S.; Zoll, S.; Morand, Y.; Descombes, S.; Guillaumot, B.; Hobbs, C.; Rochat, N.; et al. Atomic Vapour Deposition (AVD™) Process for High Performance HfO$_2$ Dielectric Layers. *MRS Proc.* **2004**, *811*, 287–292. [CrossRef]
8. Schumacher, M.; Baumann, P.K.; Seidel, T. AVD and ALD as Two Complementary Technology Solutions for Next Generation Dielectric and Conductive Thin-Film Processing. *Chem. Vapor Depos.* **2006**, *12*, 99–108. [CrossRef]
9. Leu, C.-C.; Lin, C.-H.; Chien, C.-H.; Yang, M.-J. Effects of HfO$_2$ buffer layer thickness on the properties of Pt/SrBi$_2$Ta$_2$O$_9$/HfO$_2$/Si structure. *J. Mater. Res.* **2008**, *23*, 2023–2032. [CrossRef]
10. Lin, C.-P.; Tsui, B.-Y.; Yang, M.-J.; Huang, R.-H.; Chien, C.-H. High-performance poly-silicon TFTs using HfO$_2$ gate dielectric. *IEEE Electron Device Lett.* **2006**, *27*, 360–363. [CrossRef]

11. Manikanthababu, N.; Dhanunjaya, M.; Nageswara Rao, S.V.S.; Pathak, A.P. SHI induced effects on the electrical and optical properties of HfO$_2$ thin films deposited by RF sputtering. *Nucl. Instrum. Methods Phys. Res. Sect. B Beam Interact. Mater. Atoms* **2016**, *379*, 230–234. [CrossRef]
12. He, G.; Jiang, S.; Li, W.; Zheng, C.; He, H.; Li, J.; Sun, Z.; Liu, Y.; Liu, M. Interface chemistry and electronic structure of ALD-derived HfAlO/Ge gate stacks revealed by X-ray photoelectron spectroscopy. *J. Alloys Compd.* **2017**, *716*, 1–6. [CrossRef]
13. Cho, Y.J.; Nguyen, N.V.; Richter, C.A.; Ehrstein, J.R.; Lee, B.H.; Lee, J.C. Spectroscopic ellipsometry characterization of high-k dielectric HfO$_2$ thin films and the high-temperature annealing effects on their optical properties. *Appl. Phys. Lett.* **2002**, *80*, 1249–1251. [CrossRef]
14. Takeuchi, H.; Ha, D.; King, T.-J. Observation of bulk HfO$_2$ defects by spectroscopic ellipsometry. *J. Vac. Sci. Technol. A Vac. Surf. Films* **2004**, *22*, 1337–1341. [CrossRef]
15. Buiu, O.; Lu, Y.; Mitrovic, I.Z.; Hall, S.; Chalker, P.; Potter, R.J. Spectroellipsometric assessment of HfO$_2$ thin films. *Thin Solid Films* **2006**, *515*, 623–626. [CrossRef]
16. Sancho-Parramon, J.; Modreanu, M.; Bosch, S.; Stchakovsky, M. Optical characterization of HfO$_2$ by spectroscopic ellipsometry: Dispersion models and direct data inversion. *Thin Solid Films* **2008**, *516*, 7990–7995. [CrossRef]
17. Pang, H.; Deng, N. A Forming-Free Bipolar Resistive Switching in HfO$_x$-Based Memory with a Thin Ti Cap. *Chin. Phys. Lett.* **2014**, *31*, 107303. [CrossRef]
18. Sharath, S.U.; Bertaud, T.; Kurian, J.; Hildebrandt, E.; Walczyk, C.; Calka, P.; Zaumseil, P.; Sowinska, M.; Walczyk, D.; Gloskovskii, A.; et al. Towards forming-free resistive switching in oxygen engineered HfO$_{2-x}$. *Appl. Phys. Lett.* **2014**, *104*, 063502. [CrossRef]
19. Tan, T.; Guo, T.; Wu, Z.; Liu, Z. Charge transport and bipolar switching mechanism in a Cu/HfO$_2$/Pt resistive switching cell. *Chin. Phys. B* **2016**, *25*, 117306. [CrossRef]
20. Zhang, W.; Kong, J.-Z.; Cao, Z.-Y.; Li, A.-D.; Wang, L.-G.; Zhu, L.; Li, X.; Cao, Y.-Q.; Wu, D. Bipolar Resistive Switching Characteristics of HfO$_2$/TiO$_2$/HfO$_2$ Trilayer-Structure RRAM Devices on Pt and TiN-Coated Substrates Fabricated by Atomic Layer Deposition. *Nanoscale Res. Lett.* **2017**, *12*. [CrossRef] [PubMed]
21. Kondaiah, P.; Shaik, H.; Mohan Rao, G. Studies on RF magnetron sputtered HfO$_2$ thin films for microelectronic applications. *Electron. Mater. Lett.* **2015**, *11*, 592–600. [CrossRef]
22. Bagus, P.S.; Illas, F.; Pacchioni, G.; Parmigiani, F. Mechanisms responsible for chemical shifts of core-level binding energies and their relationship to chemical bonding. *J. Electron Spectrosc. Relat. Phenom.* **1999**, *100*, 215–236. [CrossRef]
23. Kim, M.-S.; Ko, Y.-D.; Yun, M.; Hong, J.-H.; Jeong, M.-C.; Myoung, J.-M.; Yun, I. Characterization and process effects of HfO$_2$ thin films grown by metal-organic molecular beam epitaxy. *Mater. Sci. Eng. B* **2005**, *123*, 20–30. [CrossRef]
24. Guo, T.; Tan, T.; Liu, Z. Resistive switching behavior of HfO$_2$ film with different Ti doping concentrations. *J. Phys. D Appl. Phys.* **2016**, *49*, 045103. [CrossRef]
25. Miyazaki, S.; Narasaki, M.; Ogasawara, M.; Hirose, M. Characterization of ultrathin zirconium oxide films on silicon using photoelectron spectroscopy. *Microelectron. Eng.* **2001**, *59*, 373–378. [CrossRef]
26. David, D.; Godet, C. Derivation of dielectric function and inelastic mean free path from photoelectron energy-loss spectra of amorphous carbon surfaces. *Appl. Surf. Sci.* **2016**, *387*, 1125–1139. [CrossRef]
27. Mondal, S.; Chen, H.-Y.; Her, J.-L.; Ko, F.-H.; Pan, T.-M. Effect of Ti doping concentration on resistive switching behaviors of Yb$_2$O$_3$ memory cell. *Appl. Phys. Lett.* **2012**, *101*, 083506. [CrossRef]
28. Lee, M.J.; Park, Y.; Ahn, S.E.; Kang, B.S.; Lee, C.B.; Kim, K.H.; Xianyu, W.X.; Yoo, I.K.; Lee, J.H.; Chung, S.J.; et al. Comparative structural and electrical analysis of NiO and Ti doped NiO as materials for resistance random access memory. *J. Appl. Phys.* **2008**, *103*, 013706. [CrossRef]
29. Tang, T.; Zhang, Z.M.; Ding, Z.J.; Yoshikawa, H. Deriving Effective Energy Loss Function for Silver from XPS Spectrum. *Phys. Procedia* **2012**, *32*, 165–172. [CrossRef]
30. Miyazaki, S. Photoemission study of energy-band alignments and gap-state density distributions for high-k gate dielectrics. *J. Vac. Sci. Technol. B Microelectron. Nanometer Struct.* **2001**, *19*, 2212. [CrossRef]
31. Miyazaki, S.; Narasaki, M.; Ogasawara, M.; Hirose, M. Chemical and electronic structure of ultrathin zirconium oxide films on silicon as determined by photoelectron spectroscopy. *Solid-State Electron.* **2002**, *46*, 1679–1685. [CrossRef]

32. Xu, D.-P.; Yu, L.-J.; Chen, X.-D.; Chen, L.; Sun, Q.-Q.; Zhu, H.; Lu, H.-L.; Zhou, P.; Ding, S.-J.; Zhang, D.W. In Situ Analysis of Oxygen Vacancies and Band Alignment in HfO$_2$/TiN Structure for CMOS Applications. *Nanoscale Res. Lett.* **2017**, *12*, 311. [CrossRef] [PubMed]
33. Huang, M.L.; Chang, Y.C.; Chang, Y.H.; Lin, T.D.; Kwo, J.; Hong, M. Energy-band parameters of atomic layer deposited Al$_2$O$_3$ and HfO$_2$ on In$_x$Ga$_{1-x}$As. *Appl. Phys. Lett.* **2009**, *94*, 052106. [CrossRef]
34. Gaumer, C.; Martinez, E.; Lhostis, S.; Guittet, M.-J.; Gros-Jean, M.; Barnes, J.-P.; Licitra, C.; Rochat, N.; Barrett, N.; Bertin, F.; et al. Impact of the TiN electrode deposition on the HfO$_2$ band gap for advanced MOSFET gate stacks. *Microelectron. Eng.* **2011**, *88*, 72–75. [CrossRef]
35. Martínez, F.L.; Toledano-Luque, M.; Gandía, J.J.; Cárabe, J.; Bohne, W.; Röhrich, J.; Strub, E.; Mártil, I. Optical properties and structure of HfO$_2$ thin films grown by high pressure reactive sputtering. *J. Phys. D Appl. Phys.* **2007**, *40*, 5256–5265. [CrossRef]
36. Cantas, A.; Aygun, G.; Basa, D.K. In-situ spectroscopic ellipsometry and structural study of HfO$_2$ thin films deposited by radio frequency magnetron sputtering. *J. Appl. Phys.* **2014**, *116*, 083517. [CrossRef]
37. Vargas, M.; Murphy, N.R.; Ramana, C.V. Structure and optical properties of nanocrystalline hafnium oxide thin films. *Opt. Mater.* **2014**, *37*, 621–628. [CrossRef]
38. Blanchin, M.-G.; Canut, B.; Lambert, Y.; Teodorescu, V.S.; Barău, A.; Zaharescu, M. Structure and dielectric properties of HfO$_2$ films prepared by a sol-gel route. *J. Sol-Gel Sci. Technol.* **2008**, *47*, 165–172. [CrossRef]
39. Xue, K.-H.; Blaise, P.; Fonseca, L.R.C.; Nishi, Y. Prediction of Semimetallic Tetragonal Hf$_2$O$_3$ and Zr$_2$O$_3$ from First Principles. *Phys. Rev. Lett.* **2013**, *110*, 065502. [CrossRef] [PubMed]
40. Martin, N.; Rousselot, C.; Rondot, D.; Palmino, F.; Mercier, R. Microstructure modification of amorphous titanium oxide thin films during annealing treatment. *Thin Solid Films* **1997**, *300*, 113–121. [CrossRef]
41. Aguirre, B.; Vemuri, R.S.; Zubia, D.; Engelhard, M.H.; Shutthananadan, V.; Bharathi, K.K.; Ramana, C.V. Growth, microstructure and electrical properties of sputter-deposited hafnium oxide (HfO$_2$) thin films grown using a HfO$_2$ ceramic target. *Appl. Surf. Sci.* **2011**, *257*, 2197–2202. [CrossRef]
42. Ramzan, M.; Rana, A.M.; Ahmed, E.; Wasiq, M.F.; Bhatti, A.S.; Hafeez, M.; Ali, A.; Nadeem, M.Y. Optical characterization of hafnium oxide thin films for heat mirrors. *Mater. Sci. Semicond. Process.* **2015**, *32*, 22–30. [CrossRef]
43. Pal, A.; Narasimhan, V.K.; Weeks, S.; Littau, K.; Pramanik, D.; Chiang, T. Enhancing ferroelectricity in dopant-free hafnium oxide. *Appl. Phys. Lett.* **2017**, *110*, 022903. [CrossRef]
44. Polakowski, P.; Müller, J. Ferroelectricity in undoped hafnium oxide. *Appl. Phys. Lett.* **2015**, *106*, 232905. [CrossRef]
45. Park, M.H.; Lee, Y.H.; Kim, H.J.; Kim, Y.J.; Moon, T.; Kim, K.D.; Müller, J.; Kersch, A.; Schroeder, U.; Mikolajick, T.; et al. Ferroelectricity and Antiferroelectricity of Doped Thin HfO$_2$-Based Films. *Adv. Mater.* **2015**, *27*, 1811–1831. [CrossRef] [PubMed]
46. Kim, K.D.; Park, M.H.; Kim, H.J.; Kim, Y.J.; Moon, T.; Lee, Y.H.; Hyun, S.D.; Gwon, T.; Hwang, C.S. Ferroelectricity in undoped-HfO$_2$ thin films induced by deposition temperature control during atomic layer deposition. *J. Mater. Chem. C* **2016**, *4*, 6864–6872. [CrossRef]
47. Fu, W.-E.; Chang, Y.-Q.; Chen, Y.-C.; Secula, E.M.; Seiler, D.G.; Khosla, R.P.; Herr, D.; Michael Garner, C.; McDonald, R.; Diebold, A.C. Post-deposition annealing analysis for HfO$_2$ thin films using GIXRR/GIXRD. In *AIP Conference Proceedings*; American Institute of Physics: College Park, MD, USA, 2009; Volume 1173, pp. 122–126.
48. Pandey, S.; Kothari, P.; Sharma, S.K.; Verma, S.; Rangra, K.J. Impact of post deposition annealing in O$_2$ ambient on structural properties of nanocrystalline hafnium oxide thin film. *J. Mater. Sci. Mater. Electron.* **2016**, *27*, 7055–7061. [CrossRef]
49. Ramadoss, A.; Krishnamoorthy, K.; Kim, S.J. Facile synthesis of hafnium oxide nanoparticles via precipitation method. *Mater. Lett.* **2012**, *75*, 215–217. [CrossRef]
50. Ayyub, P.; Palkar, V.R.; Chattopadhyay, S.; Multani, M. Effect of crystal size reduction on lattice symmetry and cooperative properties. *Phys. Rev. B* **1995**, *51*, 6135–6138. [CrossRef]
51. Matovic, B.; Pantic, J.; Lukovic, J.; Cebela, M.; Dmitrovic, S.; Mirkovic, M.; Prekajski, M. A novel reduction–oxidation synthetic route for hafnia. *Ceram. Int.* **2016**, *42*, 615–620. [CrossRef]
52. Chen, S.; Li, Q.; Ferguson, I.; Lin, T.; Wan, L.; Feng, Z.C.; Zhu, L.; Ye, Z. Spectroscopic ellipsometry studies on ZnCdO thin films with different Cd concentrations grown by pulsed laser deposition. *Appl. Surf. Sci.* **2017**, *421*, 383–388. [CrossRef]

53. Liu, Y.; Li, Q.X.; Wan, L.Y.; Kucukgok, B.; Ghafari, E.; Ferguson, I.T.; Zhang, X.; Wang, S.; Feng, Z.C.; Lu, N. Composition and temperature dependent optical properties of Al_xGa_{1-x} N alloy by spectroscopic ellipsometry. *Appl. Surf. Sci.* **2017**, *421*, 389–396. [CrossRef]
54. Ding, L.; Friedrich, M.; Fronk, M.; Gordan, O.D.; Zahn, D.R.T.; Chen, L.; Wei Zhang, D.; Cobet, C.; Esser, N. Correlation of band gap position with composition in high-k films. *J. Vac. Sci. Technol. B Nanotechnol. Microelectron. Mater. Process. Meas. Phenom.* **2014**, *32*, 03D115. [CrossRef]
55. Di, M.; Bersch, E.; Diebold, A.C.; Consiglio, S.; Clark, R.D.; Leusink, G.J.; Kaack, T. Comparison of methods to determine bandgaps of ultrathin HfO_2 films using spectroscopic ellipsometry. *J. Vac. Sci. Technol. A Vac. Surf. Films* **2011**, *29*, 041001. [CrossRef]
56. Park, J.-W.; Lee, D.-K.; Lim, D.; Lee, H.; Choi, S.-H. Optical properties of thermally annealed hafnium oxide and their correlation with structural change. *J. Appl. Phys.* **2008**, *104*, 033521. [CrossRef]

© 2018 by the authors. Licensee MDPI, Basel, Switzerland. This article is an open access article distributed under the terms and conditions of the Creative Commons Attribution (CC BY) license (http://creativecommons.org/licenses/by/4.0/).

Article

Effect of Lithium Doping on Microstructural and Optical Properties of ZnO Nanocrystalline Films Prepared by the Sol-Gel Method

Hung-Pin Hsu [1], Der-Yuh Lin [2,*], Cheng-Ying Lu [2], Tsung-Shine Ko [2] and Hone-Zern Chen [3]

1. Department of Electronic Engineering, Ming Chi University of Technology, No. 84 Gongzhuan Road, New Taipei City 24301, Taiwan; hphsu@mail.mcut.edu.tw
2. Department of Electronic Engineering, National Changhua University of Education, No. 2, Shi-Da Road, Changhua 50074, Taiwan; M0453001@mail.ncue.edu.tw (C.-Y.L.); tsko@cc.ncue.edu.tw (T.-S.K.)
3. Department of Electronic Engineering, Hsiuping University of Science and Technology, No.11 Gongye Road, Taichung 41280, Taiwan; hzc@mail.hust.edu.tw
* Correspondence: dylin@cc.ncue.edu.tw

Received: 28 April 2018; Accepted: 16 May 2018; Published: 19 May 2018

Abstract: The $Zn_{1-x}Li_xO$ (x = 0, 0.01, 0.03, and 0.05) nanocrystalline films were synthesized on silicon (Si) substrates by using the sol-gel method. The crystal structure and surface morphology of these films were investigated by X-ray diffraction (XRD) and field emission scanning electron microscopy (FE-SEM). We observed that the average grain size was gradually reduced with increasing doping Li content. Photoluminescence (PL) spectra show that increasing the Li content will deteriorate the crystalline quality and result in the decrease of ultraviolet emission from the excitonic recombination and the enhancement of visible emission from the recombination between the intrinsic defects. The current-voltage properties of $Zn_{1-x}Li_xO$ nanocrystalline films were also studied under dark and photo-illumination for photo-detection applications. The normalized photo-to-dark-current ratio $(I_{photo} - I_{dark})/I_{dark}$ has been enhanced from 315 to 4161 by increasing the Li content of the $Zn_{1-x}Li_xO$ nanocrystalline films from zero to 0.05.

Keywords: ZnO; Sol-Gel methed; nanocrystalline

1. Introduction

Wide band gap materials with transparent properties are important for photovoltaic and optoelectronic devices. Zinc oxide (ZnO) is one of the attractive transparent compound semiconductor materials with wide band gap (3.3 eV) and a large exciton binding energy (60 meV) at room temperature [1,2]. Due to its specific properties, ZnO-based optoelectronic devices are believed to be promising candidates for transparent electronics applications [3], such as in thin film transistors [4–6], solar cells [7,8], photodetectors, [9] and light emitting devices [10]. Furthermore, due to its high electromechanical coupling coefficients, ZnO can be used in surface acoustic wave devices and chemical biosensors [11]. ZnO crystals or films have been successfully prepared by various kinds of fabrication methods, such as Radio Frequency (RF) magnetic sputtering, chemical vapor deposition, pulsed laser deposition, and sol-gel methods. Those fabricating systems require high vacuum and complicated temperature control processes, while the sol-gel process offers more merits due to ease of control of the chemical composition and a much simpler method for large-area coating at a low cost [12]. To realize an electronic device consisting of *p*- and/or *n*-type characteristics, the doping technique is an important issue. Until now, several studies on ZnO doped with group III elements [13–15], transition metals [16,17], and other elements [18,19] have been explored. The dopants may not only change the dominant crystal orientation, but also play an important role in the enhancement of the photo-to-dark-current ratio via the doping-induced defect states [20]. However, it is difficult to achieve good and reproducible p-type ZnO due to low solubility of the dopant and high self-compensation [21]. The p-type ZnO was reported by doping Li since the solubility of Li in ZnO can be up to 30% [22]. The interest in using Li as the dopant in ZnO is based on its potential ability to act as a p-type dopant and ferroelectricity behavior as reported in II–VI semiconductors [23]. Due to its specific and interesting properties, further study of the microstructural and optical properties of lithium doping on of ZnO is not only thought-provoking, but also important.

In this paper, we report a study of the doping effect of Li atoms on the optical and structural properties of $Zn_{1-x}Li_xO$ (x = 0, 0.01, 0.03, and 0.05) nanocrystalline films. The X-ray diffraction (XRD) and field emission scanning electron microscopy (FESEM) were used to characterize the crystal structure and surface morphology of $Zn_{1-x}Li_xO$ nanocrystalline films. The emission behaviors were probed by the photoluminescence (PL) technique. In addition, we also carried out the current-voltage (I–V) measurements under dark and UV illumination to investigate their optical detection properties.

2. Experimental

The source solutions for the growth of $Zn_{1-x}Li_xO$ nanocrystalline films were prepared with the precursors of zinc acetate dihydrate $(CH_3COO)_2$ $Zn·2H_2O$ and lithium acetate dehydrate $C_2H_3LiO_2·2H_2O$ which were dissolved in stoichiometric proportions in deionized water. The concentration of metal ions was kept at 0.5 M with lithium mole ratios of 0%, 1%, 2%, 3%, and 5%, respectively. We added ethanolamine into the solutions to obtain stable precursor solutions. After stirring at 150 °C for 1 h on a hotplate, we can obtain transparent solutions for crystal growth. A standard substrate cleaning process using ethanol (95% purity), acetone (99.87% purity), isopropyl alcohol (99.9% purity), hydrochloric acid (36% purity) and deionized water was done before crystal growth. Then, the Si substrates were rinsed in deionized water and dried by flowing nitrogen gas. Each layer was grown on the Si substrate by a spin coater with three steps: the first step with spinning rate = 100 rpm for 10 s, then the second step with a spinning rate of 3000 rpm for 30 s, then setting the preheating temperature at 300 °C for 2 mins. Ten layers were stacked on Si substrates. Then, the samples were annealed at 600 °C for 2 min by rapid thermal annealing (RTA) treatment with a heating rate of 600 °C/min.

The XRD patterns were studied by using Rigaku D/max-2200 X-ray diffractometer with Cu-Kα radiation. The surface morphology and cross-sectional views of $Zn_{1-x}Li_xO$ nanocrystalline films were investigated by field emission scanning electron microscopy (FE-SEM, HITACHI S-4800) at 3.0 kV. For PL measurements, room temperature photoluminescence (RTPL) spectroscopy was used to measure optical emissions by an He-Cd laser with a wavelength of 325 nm. In I–V characteristics, the samples were measured under dark and UV lamp illumination (365 nm) by an HP 4145 semiconductor parameter analyzer in an applied voltage from -5 to $+5$ V.

3. Results and Discussion

Figure 1 depicts the XRD patterns of $Zn_{1-x}Li_xO$ nanocrystalline films with different Li contents from x = 0 to 0.05. It is clear from observations that only one main peak located at 2θ = 34.56 degree, which demonstrates that all the $Zn_{1-x}Li_xO$ nanocrystalline films consisted of a unique phase. The $Zn_{1-x}Li_xO$ nanocrystalline film is well crystallized in a wurtzite structure with a (002) preferred orientation in the direction parallel to the c-axis. The lattice constant of c-axis can be extracted to be 5.2069 Å. E. Nurfani et al. have reported that Ti dopants change the dominant crystal orientation from (002) to (103), and also slightly extend the c-axis of the ZnO lattice parameter [20]. The results obtained in this work show that the doping of Li atoms does not cause obvious structure changes or lattice extensions. This might be due to the fact that the radii of Ti^{2+} or Ti^{4+} ions are larger than those of Li^+ ions, because the atomic number of Ti is 22 while that of Li is only 3. The inset presents the full width at half maximum (FWHM) values of the XRD peaks in a function of Li content. It is noticed that the intensity of the (002) peaks gradually decreases with increasing Li content, at the same time, the FWHM are constantly broadened. We also deduced the crystallite size of $Zn_{1-x}Li_xO$ nanocrystalline films from the Debye–Scherrer's equation [24,25] $D = k\lambda/\beta \cos\theta$, where k is the Scherer constant (k = 0.9), λ is the wavelength of the X-ray radiation (0.154 nm), β is the FWHM in radians, and θ is the Bragg diffraction angle. The results are 39.8, 36.9, 35.6, and 34.6 nm for x = 0, 0.01, 0.03, 0.05, respectively. The crystallite size is decreasing, which is due to the crystalline quality deteriorating by increasing the doping concentration of the Li atoms.

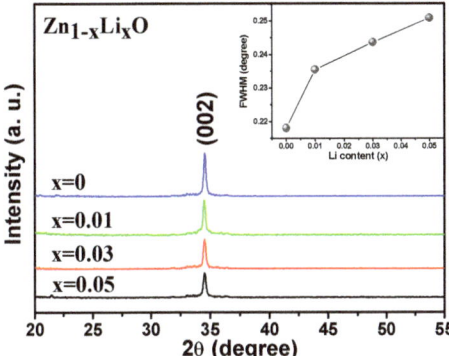

Figure 1. XRD patterns of $Zn_{1-x}Li_xO$ nanocrystalline films. The inset shows that FWHM broadens with various Li content.

The grain size of $Zn_{1-x}Li_xO$ nanocrystalline films can also be observed by FE-SEM. Figure 2 shows both the surface morphology and cross-sectional views of $Zn_{1-x}Li_xO$ nanocrystalline films. From the top view of $Zn_{1-x}Li_xO$ nanocrystalline surface morphology, the average grain size is determined to be 88.02 to 71.16 nm for x = 0 to 0.05, respectively, and it is observed that the average grain size is gradually reduced with increasing Li doping concentration. The tendency of grain change further confirms the observation in the XRD patterns. It is noted that the crystallite size is supposed to be the size of a coherently diffracting domain and is not exactly to be the same as the particle size [26] Furthermore, it has been found that the XRD peak can be widened by defects or internal stress, so the mean crystallite size calculated by the Debye–Scherrer's equation is smaller than the actual value [27]. In the side views, we can observe that the films are quite uniform with thicknesses of about 100 nm for all samples, and the pictures further confirm the c-axis orientation with columnar grains running perpendicular to the substrate.

Figure 3 presents the PL spectra of $Zn_{1-x}Li_xO$ nanocrystalline films for x = 0 to 0.05 taken at room temperature. The PL spectra consist of an ultraviolet (UV) emission peak and a weak green yellow visible (VIS) emission band. The sharp UV emission peaks resulted from the exciton recombination [28,29]. It is obvious that the PL intensity decreases with increasing Li doping concentration, which indicates that the doping Li atoms will deteriorate the crystalline quality. All the UV emission peaks seem located around 3.24 eV, and in careful observation we can find a slight blue shift with increasing Li doping concentration. The broader VIS emission below 2.4 eV with a peak around 2.1 eV reflects the characteristic luminescence associated with singly ionized defects, such as oxygen vacancy or doping-induced defects [30,31] Comparing the intensity of the UV emission to the VIS emission, we observe the ratio of I_{UV}/I_{VIS} is 8.68, 4.21, 2.52, and 1.54 for various Li doping concentrations (x = 0, 0.01, 0.03, and 0.05, respectively). We believe that the doping Li atoms are responsible for the decreasing ratio of I_{UV}/I_{VIS}.

Furthermore, we study the optical properties of these thin films by means of I–V curves under dark and UV illumination. The photocurrent was measured by a 30-W Xe lamp with the incident wavelength of 365 nm as the irradiation source. In Figure 4, the I–V curves of $Zn_{1-x}Li_xO$ nanocrystalline films were measured under dark and UV illumination in the voltage range from −5 to +5 V. For optical detection applications, the relative optical response can be defined by the normalized photo-to-dark-current ratio $(I_{photo} - I_{dark})/I_{dark}$. The results at +5 V are calculated and listed in Table 1. This shows that the doping Li atoms are helpful to enhance the ratio of $(I_{photo} - I_{dark})/I_{dark}$, which has been raised more than ten times from 315 to 4161.

Figure 2. FE-SEM surface morphology (**a–d**) and cross sectional view images (**e–h**) of $Zn_{1-x}Li_xO$ nanocrystalline films with x = 0 to 0.05. The average grain size is determined to be 88.02 to 71.16 nm for x = 0 to 0.05.

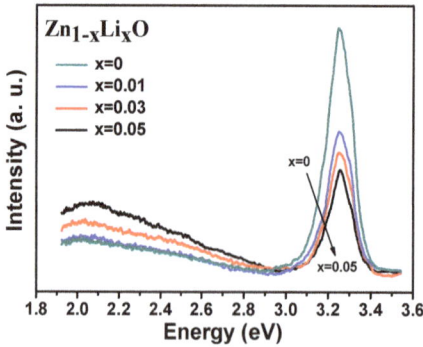

Figure 3. PL spectra of $Zn_{1-x}Li_xO$ nanocrystalline films with x = 0 to 0.05.

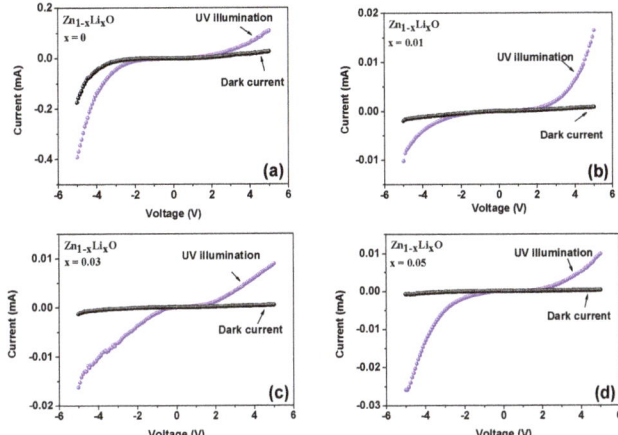

Figure 4. I–V characteristics of $Zn_{1-x}Li_xO$ nanocrystalline films [(**a**) x = 0, (**b**) x= 0.01, (**c**) x=0.03, (**d**) x= 0.05] measured under dark and photo-illumination in the voltage from −5 V to +5 V.

Table 1. Values of I_{photo}, I_{dark}, and $(I_{photo} - I_{dark})/I_{dark}$ at +5 V of $Zn_{1-x}Li_xO$ nanocrystalline films.

$Zn_{1-x}Li_xO$ (x)	I_{photo} (mA)	I_{dark} (μA)	$(I_{photo} - I_{dark})/I_{dark}$ (%)
0	0.1084	26.12	315
0.01	0.0164	0.79	1975
0.03	0.0089	0.41	2071
0.05	0.0098	0.23	4161

4. Conclusions

The ZnO doped with Li (0–5%) was successfully synthesized by the sol-gel method. XRD patterns showed a single crystalline phase of $Zn_{1-x}Li_xO$. The composition-dependent crystallite size was deduced by Debye–Scherrer's equation, and further confirmed by the surface morphology from FE-SEM images. The PL spectra show a slight blue shift with an increase in Li content. The intensity of UV emission from exciton recombination decreases with increasing Li content, which is due to crystalline deterioration. The higher ratio of the normalized photo-to-dark-current ratio resulting from the doping of Li atoms demonstrates their potential in photodetector applications.

Author Contributions: H.P.H. and D.Y.L. conceived and designed the experiments. C.Y.L. and H.Z.C. prepared the materials. C.Y.L. and T.S.K. performed the experiments. H.P.H., D.Y.L., and T.S.K. analyzed data. H.P.H., D.Y.L., and H.Z.C. wrote the manuscript.

Acknowledgments: This work was financial supported by Ministry of Science and Technology of Taiwan under grant no. MOST 105-2112-M-018-006.

Conflicts of Interest: The authors declare no conflict of interest.

References

1. Özgür, Ü.; Alivov, Y.I.; Liu, C.; Teke, A.; Reshchikov, M.A.; Dogan, S.; Avrutin, V.; Cho, S.J.; Morkoç, H. A comprehensive review of ZnO materials and devices. *J. Appl. Phys.* **2008**, *98*, 041301. [CrossRef]
2. Özgür, Ü.; Hofstetter, D.; Morkoç, H. ZnO devices and applications: A review of current status and future prospects. *Proc. IEEE* **2010**, *98*, 1255–1268. [CrossRef]
3. Wager, J.F. Transparent electronics. *Science* **2003**, *300*, 1245–1246. [CrossRef] [PubMed]
4. Oh, B.; Kim, Y.; Lee, H.; Kim, B.; Park, H.; Han, J.; Heo, G.; Kim, T.; Kim, K.; Seo, D. High-performance ZnO thin-film transistor fabricated by atomic layer deposition. *Semicond. Sci. Technol.* **2011**, *26*, 085007. [CrossRef]

5. Hirao, T.; Furuta, M.; Hiramatsu, T.; Matsuda, T.; Li, C.; Furuta, H.; Hokari, H.; Yoshida, M.; Ishii, H.; Kakegawa, M. Bottom-Gate Zinc Oxide Thin-Film Transistors (ZnO TFTs) for AM-LCDs. *IEEE Trans. Electron Devices* **2008**, *55*, 3136–3142. [CrossRef]
6. Fortunato, E.M.C.; Barquinha, P.M.C.; Pimentel, A.C.M.B.G.; Gonçalves, A.M.F.; Marques, A.J.S.; Martins, R.F.P.; Pereira, L.M.N. Wide-bandgap high-mobility ZnO thin-film transistors produced at room temperature. *Appl. Phys. Lett.* **2004**, *85*, 2541–2543. [CrossRef]
7. Pietruszka, R.; Witkowski, B.S.; Gieraltowska, S.; Caban, P.; Wachnicki, L.; Zielony, E.; Gwozdz, K.; Bieganski, P.; Placzek-Popko, E.; et al. New efficient solar cell structures based on zinc oxide nanorods. *Sol. Energy Mater. Sol. Cells* **2015**, *143*, 99–104. [CrossRef]
8. Vittal, R.; Ho, K.C. Zinc oxide based dye-sensitized solar cells: A review. *Renew. Sustain. Energy Rev.* **2017**, *70*, 920–935. [CrossRef]
9. Kind, H.; Yan, H.; Messer, B.; Law, M.; Yang, P. Nanowire ultraviolet photodetectors and optical switches. *Adv. Mater.* **2002**, *14*, 158–160. [CrossRef]
10. Liu, C.H.; Zapien, J.A.; Yao, Y.; Meng, X.M.; Lee, C.S.; Fan, S.S.; Lifshitz, Y.; Lee, S.T. High-Density, ordered ultraviolet light-emitting ZnO nanowire arrays. *Adv. Mater.* **2003**, *15*, 838–841. [CrossRef]
11. Du, X.Y.; Fu, Y.Q.; Tan, S.C.; Luo, J.K.; Flewitt, A.J.; Maeng, S.; Kim, S.H.; Choi, Y.J.; Lee, D.S.; Park, N.M.; et al. ZnO film for application in surface acoustic wave device. *J. Phys. Conf. Ser.* **2007**, *76*, 012035. [CrossRef]
12. Znaidi, L. Sol-gel-deposited ZnO thin films: A review. *Mater. Sci. Eng. B* **2010**, *174*, 18–30. [CrossRef]
13. Huang, Q.; Wang, Y.; Wang, S.; Zhang, D.; Zhao, Y.; Zhang, X. Transparent conductive ZnO:B films deposited by magnetron sputtering. *Thin Solid Films* **2012**, *520*, 5960–5964. [CrossRef]
14. Geng, Y.; Guo, L.; Xu, S.S.; Sun, Q.Q.; Ding, S.J.; Lu, H.L.; Zhang, D.W. Influence of Al doping on the properties of ZnO thin films grown by atomic layer deposition. *J. Phys. Chem. C* **2011**, *115*, 12317–12321. [CrossRef]
15. Zhao, J.L.; Sun, X.W.; Ryu, H.; Moon, Y.B. Thermally stable transparent conducting and highly infrared reflective Ga-doped ZnO thin films by metal organic chemical vapor deposition. *Opt. Mater.* **2011**, *33*, 768–772. [CrossRef]
16. Tsay, C.Y.; Cheng, H.C.; Tung, Y.T.; Tuan, W.H.; Lin, C.K. Effect of Sn-doped on microstructural and optical properties of ZnO thin films deposited by sol-gel method. *Thin Solid Films* **2008**, *517*, 1032–1036. [CrossRef]
17. Lin, C.C.; Young, S.L.; Kung, C.Y.; Jhang, M.C.; Lin, C.H.; Kao, M.C.; Chen, H.Z.; Ou, C.R.; Cheng, C.C.; Lin, H.H. Effect of Fe doping on the microstructure and electrical properties of transparent ZnO nanocrystalline films. *Thin Solid Films* **2013**, *529*, 479–482. [CrossRef]
18. Bhachu, D.S.; Sankar, G.; Parkin, I.P. Aerosol assisted chemical vapor deposition of transparent conductive zinc oxide films. *Chem. Mater.* **2012**, *24*, 4704–4710. [CrossRef]
19. Young, S.L.; Kao, M.C.; Chen, H.Z.; Shih, N.F.; Kung, C.Y.; Chen, C.H. Mg doping effect on the microstructural and optical properties of ZnO nanocrystalline films. *J. Nanomater.* **2015**, *2015*, 627650. [CrossRef]
20. Nurfani, E.; Zuhairah, N.; Kurniawan, R.; Muhammady, S.; Sutjahja, I.M.; Winata, T.; Darma, Y. Infulence of Ti doping on the performance of a ZnO-based photodetector. *Mater. Res. Express* **2017**, *4*, 024001. [CrossRef]
21. Chand, P.; Gaur, A.; Kumar, A.; Gaur, U.K. Structural, morphological and optical study of Li doped ZnO thin films on Si (100) substrate deposited by pulsed laser deposion. *Ceram. Int.* **2014**, *40*, 11915–11923. [CrossRef]
22. Lu, J.G.; Zhang, Y.Z.; Ye, Z.Z.; Zeng, Y.J.; He, H.P.; Zhu, L.P.; Huang, J.Y.; Wang, L.; Yuan, J.; Zhao, B.H.; et al. Control of p- and n-type conductivities in Li-doped ZnO thin films. *Appl. Phys. Lett.* **2006**, *89*, 112113. [CrossRef]
23. Fu, D.J.; Lee, J.C.; Choi, S.W.; Park, C.S.; Panin, G.N.; Kang, T.W. Ferroelectricity in Mn-implanted CdTe. *Appl. Phys. Lett.* **2003**, *83*, 2214–2216. [CrossRef]
24. Singh, A.K.; Viswanath, V.; Janu, V.C. Synthesis, effect of capping agents, structural, optical and photoluminescence properties of ZnO nanoparticles. *J. Lumin.* **2009**, *129*, 874–878. [CrossRef]
25. Rattana, T.; Suwanboon, S.; Amornpitoksuk, P.; Haidoux, A.; Limsuwan, P. Improvement of optical properties of nanocrystalline Fe-doped ZnO powders through precipitation method from citrate-modified zinc nitrate solution. *J. Alloys Compd.* **2009**, *480*, 603–607. [CrossRef]
26. Klug, H.P.; Alexander, L.E. X-ray diffraction procedures for polycrystalline and amorphous materials. *J. Appl. Crystallogr.* **1975**, *8*, 573–574.
27. Ashour, A.; Kaid, M.A.; El-Sayed, N.Z.; Ibrahim, A.A. Physical properties of ZnO thin films deposited by spray pyrolysis technique. *Appl. Surf. Sci.* **2006**, *252*, 7844–7848. [CrossRef]

28. Kenanakis, G.; Androulidaki, M.; Vernardou, D.; Katsarakis, N.; Koudoumas, E. Photoluminescence study of ZnO structures grown by aqueous chemical growth. *Thin Solid Films* **2011**, *520*, 1353–1357. [CrossRef]
29. Banerjee, D.; Lao, J.Y.; Wang, D.Z.; Huang, J.Y.; Steeves, D.; Kimball, B.; Ren, Z.F. Synthesis and photoluminescence studies on ZnO nanowires. *Nanotechnology* **2004**, *15*, 404–409. [CrossRef]
30. McCluskey, M.D.; Jokela, S.J. Defects in ZnO. *J. Appl. Phys.* **2009**, *106*, 071101. [CrossRef]
31. Lin, B.; Fu, Z.; Jia, Y. Green luminescent center in undoped zinc oxide films deposited on silicon substrates. *Appl. Phys. Lett.* **2001**, *79*, 943–945. [CrossRef]

© 2018 by the authors. Licensee MDPI, Basel, Switzerland. This article is an open access article distributed under the terms and conditions of the Creative Commons Attribution (CC BY) license (http://creativecommons.org/licenses/by/4.0/).

Article

Epitaxial Crystallization of Precisely Methyl-Substituted Polyethylene Induced by Carbon Nanotubes and Graphene

Weijun Miao [1], Yiguo Li [2], Libin Jiang [1], Feng Wu [1], Hao Zhu [1], Hongbing Chen [1] and Zongbao Wang [1,*]

1. Ningbo Key Laboratory of Specialty Polymers, Faculty of Materials Science and Chemical Engineering, Ningbo University, Ningbo 315211, China; 15988621876@163.com (W.M.); 18892615669@163.com (L.J.); 18892615653@163.com (F.W.); zh2320015@163.com (H.Z.); chenhongbing@nbu.edu.cn (H.C.)
2. Anhui Collaborative Innovation Center for Petrochemical New Materials, School of Chemistry and Chemical Engineering, Anqing Normal University, Anqing 246011, China; liyiguo@aqnu.edu.cn
* Correspondence: wangzongbao@nbu.edu.cn

Received: 9 March 2018; Accepted: 11 April 2018; Published: 16 April 2018

Abstract: How large of a substituent/branch a polyethylene possesses that can still be induced by nanofillers to form ordered chain structures is interesting, but uncertain. To solve this problem, precisely methyl-substituted polyethylene (PE21M) was chosen as a model to prepare its one-dimensional and two-dimensional nanocomposites with carbon nanotubes (CNTs) and graphene via solution crystallization. It is shown that kebab-like and rod-like nanofiller-induced crystals were separately observed on the surfaces of CNTs and graphene and the density of rod-like crystals is significantly less than kebab-like ones. The results of differential scanning calorimetry (DSC) and X-ray diffraction (XRD) reveal that CNTs and graphene cannot induce polymers with the substituent volume greater than, or equal to, 2 Å (methyl) to form ordered lattice structure, but CNTs exhibit the better nucleation effect, providing us with guidance to manipulate the physical performance of polymer composites on the basis of the size of the substituent and the type of nanofiller.

Keywords: PE21M; carbonaceous nanofiller; epitaxial crystallization

1. Introduction

Epitaxial crystallization of semi-crystalline polymers on foreign surfaces can produce controllable crystal structures and morphologies which, therefore, provides an efficient way to tailor the physical performance of polymeric materials [1–4]. It is widely recognized that the mechanism of epitaxial crystallization is based on some certain crystallographic matches [1], and the mismatching of the contact lattice planes between the substrate and polymers cannot exceed 15%. For example; one-dimensional or two-dimensional lattice matches generate special interactions between the polymer chains and the substrate in the contacting interface which can alter not only the crystal structure and morphology of polymers, but the crystallization kinetics [2–6]. One-dimensional carbon nanotubes (CNTs) and two-dimensional graphene are frequently used as substrates to induce polymer crystallization because of their high surface area. It has been well documented that the topological structure of substrates has a significant influence on the polymer crystallization behavior [7–15]. Generally, the mechanism of CNTs inducing polymers to form disk-like crystal lamellae was described as "soft epitaxy" [16–18]. While, for two-dimensional graphene, lattice matching should play the dominant role in surface-induced polymer epitaxial crystallization [19–22]

Polyethylene is one of the most studied polymers due to its versatility in a very wide range of applications. The common method for modifying the structures and properties of polyethylene is randomly copolymerizing ethylene with 1-alkene or with polar/nonpolar comonomers. However, random copolymerization leads to the products with a bivariate comonomer content-molar mass distribution, so the physical properties can vary depending on the molecular weight, molecular-weight distribution, branch identity, branch content, and branch distribution, and are also difficult to reproduce at a high level of consistency [23,24]. Accordingly, the relationship between the structure and performance of random copolyethylene is quite ambiguous. Polymers synthesized by acyclic diene metathesis polymerization (ADMET) offer the suitable model systems to study the relationship of the structure and performance which possess a well-defined primary structure. ADMET polymers have a general repeating unit, $-[(CH_2)_m\text{-}CHX]_n-$, where m is in the range of 8–74 and X is the group of various types placed at a precise distance along the PE backbone, such as halogen, alkyl group, or other functional groups [25–33]. These polymers were found to show the various crystalline phases depending on the side group X. Meanwhile, the aggregation state of precision macromolecules may be affected remarkably by a small change of the chain structure and the distribution of defects in crystallites can be exquisitely controlled by the crystallization kinetic, which is not yet feasible in classical branched polyethylene obtained via coordination catalysis [34,35]. In our previous works, the precise ADMET polyethylene with halogen atoms (F, Cl, Br) placed on each every 21st backbone carbon were chosen as models to investigate the structural change of polymer epitaxial crystallization. The results demonstrated that CNTs had almost no affection on the structure of PE21F and PE21Cl, but RGO induced the structural transformation of PE21Cl and PE21Br from a triclinic form to orthorhombic form, which generated extraordinarily high melting temperatures [36–38]. Such a significant change of the crystal structure resulting from surface induced epitaxial crystallization not only suggests that the precisely substituted polyethylenes are ideal models to explore the influence of the substituent on epitaxial crystallization, but also indicates that single-walled carbon nanotubes (SWCNTs) and reduced graphene oxide (RGO) possess different capacities to induce substituted polymers to generate ordered chain packing structures due to the different epitaxy mechanism of the two nanofillers. However, how large of a substituent/branch a polyethylene possesses that can still be induced by nanofillers to form ordered chain structures is unknown.

For solving this question, in this paper, the precision ADMET polyethylene with the methyl placed on every 21st backbone carbon (PE21M) was chosen as a model to investigate the epitaxial crystallization of polymer with a much bulkier substituent. The larger volume of methyl (2.0 Å) compared to those of F (1.47 Å) to Cl (1.75 Å) and Br (1.85 Å) results in a much more disordered crystal lattice and the lower melting temperature of PE21M, and the bulkier substituent is beneficial for exploring the capacities of carbonaceous nanofillers that have different mechanisms of epitaxial crystallization to induce nucleation and crystallization of substituted/branched polymers. Meanwhile, one-dimensional SWCNTs and two-dimensional RGO were employed as nanofillers to prepare PE21M-based nanocomposites and the epitaxial crystallization of nanocomposites was achieved via solution crystallization. The resultant crystal morphology and structures were analyzed by transmission electron microscopy (TEM) and X-ray diffraction (XRD) and the thermal behaviors were characterized by differential scanning calorimetry (DSC). Finally, the discrepancy of the density of crystals formed on two nanofillers was also discussed to compare their epitaxy ability. This study is expected to enhance the understanding of the influence of low-dimension carbonaceous nanofillers on the nucleation and chain packing structure of semi-crystalline polymers with large substituents/branches.

2. Experimental Section

2.1. Materials

Purified HiPco single-walled CNTs (SWCNTs, with an average diameter of 6 nm) were purchased from Times Nanotechnologies Inc (Chengdu Organic Chemicals Co. Ltd., Chengdu, China) and used as received. Reduced graphene oxide (RGO) was prepared by thermal exfoliation and reduction of graphene oxide (GO) [39]. The precisely substituted polyethylene with methyl placed on each every 21st backbone carbon has been studied, labelled as PE21M, where the number corresponds to the precise location of the side group in the PE backbone. The synthesis route of PE21M is shown in Scheme 1 [31]. The chemical structure was characterized by ^{1}H NMR (Supporting Information Figure S1). The molecular weight was determined by gel permeation chromatography (GPC) using an Agilent PL-GPC 220 instrument with HPLC-grade chloroform as the mobile phase at a flow rate of 1.0 mL·min^{-1} and a calibration with polystyrene standards (M_n = 12,962 g/mol, M_w/M_n = 2.17).

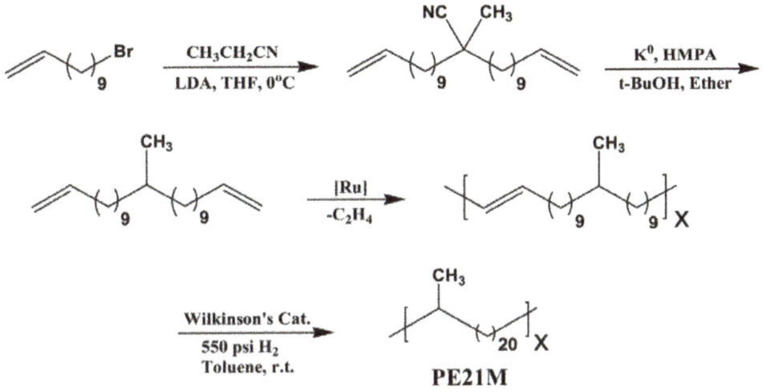

Scheme 1. Monomer and polymer synthesis process.

2.2. Sample Preparation

The fabrication of PE21M nanocomposites were described as follows. The nanofiller/p-xylene mixed solution with the nanofiller mass concentration of 0.1 wt % was sonicated for 2–3 h at 45 °C and the PE21M/p-xylene mixed solution with PE21M mass concentration of 0.1 wt % was prepared by dissolving PE21M into p-xylene at 120 °C for 2 h. Then, 10 g nanofillers/p-xylene solution was mixed with 10 g PE21M/p-xylene solution at 120 °C for 5 min. The mixture was finally quenched to the preset crystallization temperature, T_c. In order to avoid SWCNTs to agglomerate and form small bundles in solvent, the mixed solution was stirred at 1200 r·min^{-1}. The sample was isothermally filtered after crystallization for 6 h. After washing with ethanol carefully three times, nanocomposites were dried at 40 °C under vacuum for 36–48 h.

Furthermore, Qunxu's method [40] was employed to prepare PE nanocomposites by supercritical CO_2 (SC CO_2). The mixture of nanofiller/p-xylene solution and PE21M/p-xylene solution was produced by the same procedure above-mentioned. Then the mixture was quickly transferred into a stainless steel autoclave at the preset crystallization temperature T_c. SC CO_2 was then charged into the autoclave up to the desired pressure within a short time. After maintaining the supercritical fluid condition for 3 h, the system was slowly depressurized and the sample was collected and labelled.

2.3. Characterization

Nanocomposite suspensions were collected on a carbon-coated grid. The crystal morphology observation was conducted by a JEOL JEM2100 transmission electron microscope (TEM, JEOL., Tokyo, Japan) with an accelerating voltage of 200 kV. Differential scanning calorimetry (DSC) experiments were carried out using a Perkin-Elmer DSC8000 (PerkinElmer, New York, NY, USA). The samples with an average weight of 2~4 mg were heated from 30 to 100 °C at a scanning rate of 10 °C·min^{-1} under a nitrogen atmosphere. X-ray diffraction (XRD) patterns were recorded on a Bruker D8 diffractometer (Bruker, Karlsruhe, Germany), using Ni-filtered Cu Kα radiation at 40 kV and 30 mA at room temperature at an angle ranging from 5 to 40° at a rate of 3.5°·min^{-1}.

3. Results and Discussion

3.1. Morphologies of PE21M/Nanofiller Composites

3.1.1. Morphologies of PE21M/CNT Kebab-Like Crystals

As expected, the shish-kebab structures periodically form on SWCNTs at all selected experimental temperature, as exhibited in Figure 1 and the average sizes of PE21M lamellae attached to SWCNTs are listed in Table 1 on the basis of the measurement of 200 lamellae. The average diameters of the PE21M kebab crystals are about 16.3 ± 1.0, 20.2 ± 1.1, and 18.4 ± 0.9 nm for 40, 50, and 60 °C, respectively. It is evident that the average size of the PE21M kebab crystals first increases then decreases with the increase of crystallization temperature and the largest diameter of kebabs form at 50 °C. This behavior of crystal growth on SWCNTs is similar to the situations reported in PE21F, PE21Cl, and PE21Br crystals [36–38], which is attributed to the competition of nucleation and crystal growth on the CNTs. Meanwhile, stable crystal nuclei become fewer and fewer with the decrease of undercooling, which can be directly supported by the interval of kebab increasing monotonically with the increase of the crystallization temperature (Table 1). The thickness of the kebab also increases with the crystallization temperature, from 5.1 ± 0.3 nm at 40 °C to 5.6 ± 0.4 nm at 60 °C (Table 1), suggesting that the crystallization ethylene sequence is strongly dependent on the crystallization temperature. We can now conclude that the crystallization temperature has an important influence on the size and periodicity of the shish-kebab structure and the most suitable crystallization temperature for PE21M/SWCNT in p-xylene is 50 °C. Our previous studies displayed that the diameter of the PE21F kebab (54–65 nm) was smaller than that of HDPE (50–80 nm) because of the influence of substituent F [36] and the diameter of the kebab decreased rapidly from PE21F (54–65 nm) to PE21Cl (26.2 ± 1.0 nm) and PE21Br (23.2 ± 1.2 nm) with increasing substituent volume (the van der Waals radii of F, Cl, and Br are 1.47, 1.75, and 1.85 Å, respectively) [36–38]. When the van der Waals volume of the substituent increases to 2.0 Å, the diameter of the PE21M kebab decreases to 20.2 ± 1.1 nm (Table 1). The above results further confirm that the substituents act as defects in the chain which have a strong impact on the lateral growth of crystal lamellae. On the other hand, the maximum thickness of PE21M kebabs is 5.6 ± 0.4 nm, which is also the smallest compared to that of PE21F (6.4–8.6 nm), PE21Cl (6.4 ± 0.5 nm), and PE21Br (6.8 ± 0.5 nm) [36–38], further indicating that the crystallization of PE is disturbed more intensely by the larger substituent. By contrasting with our previous works, we found that the interval of PE21M kebab crystals prepared under the same supercooling degree decreases severely with the increase of the substituent's volume [36–38]. The largest interval of PE21M kebab crystals is 20.5 nm, much lower than 48.5 nm of PE21F, 57.7 nm of PE21Cl, and 34.5 nm of PE21Br [36–38]. This can be attributed to the thinner PE21M lamellae resulting in a weak repulsion between adjacent lamellae. According to the "soft epitaxy" mechanism [16–18], CNTs can induce polymers to crystallize regardless of the lattice matching between the polymer chain and the graphitic sheet. Consequently, the density of PE21M crystal nuclei developed on CNTs was almost unaffected by the bulkier substituent.

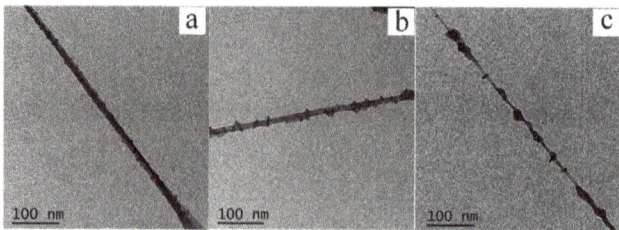

Figure 1. TEM images of PE21M/SWCNT nanocomposites produced in p-xylene at different temperatures for 6 h. (**a**) 40 °C, (**b**) 50 °C, (**c**) 60 °C (both PE21Br and SWCNT concentrations are 0.05 wt %).

Table 1. The average size of PE21M lamellae formed on SWCNT based on the TEM images of 200 lamellae.

Sample (Crystallization Temperature)	Diameter of Kebab (nm)	Thickness of Kebab (nm)	Interval of Kebab (nm)
PE21M/SWCNT (40 °C)	16.3 ± 1.0	5.1 ± 0.3	10.0 ± 0.8
PE21M/SWCNT (50 °C)	20.2 ± 1.1	5.4 ± 0.4	16.3 ± 1.0
PE21M/SWCNT (60 °C)	18.4 ± 0.9	5.6 ± 0.4	20.5 ± 1.1

The evolution of kebabs with the isothermal crystallization time was also detected. The morphologies of PE21M crystals on SWCNTs in *p*-xylene at 50 °C for different isothermal crystallization times are shown in Figure 2 and the average sizes of PE21M lamellae are summarized in Table 2. The average diameter of the PE21M kebab crystals is 10.3 ± 0.8 nm after 3 h isothermal crystallization and the continued growth during the subsequent 3 h leads to the diameter of kebabs reaching 20.2 ± 1.1 nm, i.e., the size doubled with respect to the former 3 h. The average growth rate of kebabs is about 3 nm·h^{-1} in the elapsed 6 h, which is far below that of PE21F (20 nm·h^{-1}) [36]. This suggests that the speed of PE21M chain lateral growth is limited by the bulkier substituent. According to the average diameter shown in Table 2, we can find that the diameter of kebab crystals reaches 22.4 ± 0.9 nm after 12 h isothermal crystallization, almost unchanged compared with the isothermal crystallization for 6 h. It can also be observed from Table 2 that little change happens for the thickness and periodicity of kebabs formed at the same crystallization temperatures, which further indicates that the thickness and periodicity of kebabs is directly related to the isothermal crystallization temperatures.

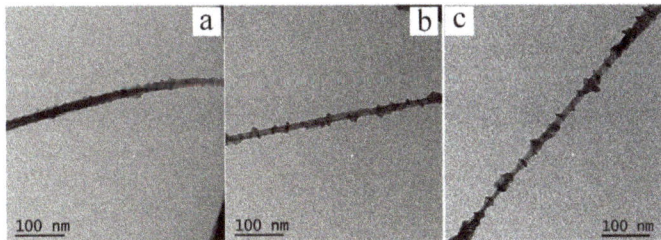

Figure 2. TEM images of PE21M/SWCNT nanocomposites produced by crystallization of PE21M on SWCNTs in *p*-xylene at 50 °C for different isothermal crystallization time. (**a**) 3 h, (**b**) 6 h, and (**c**) 12 h (both PE21Br and SWCNT concentrations are 0.05 wt %).

Table 2. The average size of PE21M lamellae formed on CNTs for different isothermal crystallization time based on the TEM images of 200 lamellae.

Crystallization Time	Diameter of Kebab (nm)	Thickness of Kebab (nm)	Interval of Kebab (nm)
3 h	10.3 ± 0.8	5.2 ± 0.3	15.8 ± 0.7
6 h	20.2 ± 1.1	5.4 ± 0.4	16.3 ± 1.0
12 h	22.4 ± 0.9	5.4 ± 0.4	15.5 ± 0.8

3.1.2. Morphologies of PE21M/RGO Rod-Like Crystals

We also investigate the effects of nanofillers with different dimensions on the crystallization behavior of PE21M. For the convenience of comparison, three crystallization temperatures (40, 50, and 60 °C) same as that of PE21M/SWCNT nanocomposites chosen to prepare PE21M/RGO nanocomposites. It can be seen from the crystals morphologies shown in Figure 3 that rod-like crystals were formed on the surface of RGO and the average sizes of PE21M crystals are listed in Table 3. Quite small particles with a size of 5.0 ± 0.2 nm can be observed on the surface of RGO nanosheets at 40 °C, which grow into the largest lamellae with an average size of 18.0 ± 1.2 nm when the crystallization temperature increases to 50 °C. This suggests that PE21M chains need more time or much higher crystallization temperature to adjust their conformations to the surface of RGO. The size of PE21M lamellae formed on RGO is 16.0 ± 1.0 nm at 60 °C, smaller than that formed at 50 °C. It is obvious that the temperature dependence of the sizes of PE21M lamellae upon the RGO surface is similar the case observed in PE21M/SWCNT composites, namely, first it increases and then decreases with increasing crystallization temperature. It has been reported that the average sizes of lamellae formed on RGO become smaller and smaller with the increase of substituent bulk as the influence of substituent on crystal growth becomes more and more serious [36–38].

In this study, the smallest average size of lamellae is observed in PE21M/RGO composites, further verifying the conclusion that substituents have a great effect on crystal growth on RGO. We can also see from Table 3 that the thickness of rod-like crystals has the same variation rule with that of kebab crystals (Table 1), i.e., it increases with increasing crystallization temperature. Moreover, the thickness of crystals formed on SWCNTs and RGO under the same crystallization condition are almost the same by comparing the data in Tables 1 and 3. These findings all suggest that the isothermal crystallization temperature plays a decisive role in the thickness of lamellae formed on RGO. By comparison, the maximal thickness of PE21M lamellae formed on RGO is about 5.5 nm, which is much smaller than 8.8 nm of PE21F, 6.5 nm of PE21Cl, and 6.9 nm of PE21Br [36–38], which is also attributed to the influence of the largest methyl substituent. Obviously, PE21M crystals formed on RGO have much poorer regularity with respect to the cases in PE21F, PE21Cl, and PE21Br crystals [36–38] and hardly arrange at 60° directions apart from each other, which indicates that the largest substituent methyl significantly disturbs the matching between the molecular chain and RGO substrate. Furthermore, it is evident from Table 3 that the density (number/0.01 μm^2) of the lamellae decreases from eight at 50 °C to five at 60 °C, which is attributed to the less stable crystal nucleus existing at a lower undercooling degree. Meanwhile, the density of PE21X (X = F, Cl, Br, and CH$_3$) crystals prepared under the same supercooling degree decreases severely with the increase of the substituent's volume [36–38]. In other words, the maximum density of PE21M crystals formed on RGO is 8/0.01 μm^2, much lower than 40/0.01 μm^2 of PE21F, 35/0.01 μm^2 of PE21Cl, and 37/0.01 μm^2 of PE21Br. Moreover, the density of PE21M crystals formed on RGO is significantly less than that formed on CNT comparing Figures 1 and 3. It is a generally accepted notion that lattice matching dominates the graphene-induced polymer epitaxial crystallization [19–22], which is of no concern for the "soft epitaxy" of CNTs [16–18]. Therefore, we can conclude that the lack of crystallographic mismatching is the main influencing factor that attaches the PE21M chain to the surface of RGO and after nucleation.

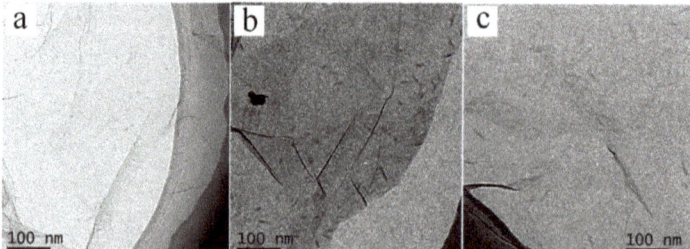

Figure 3. TEM images of PE21M/RGO nanocomposites produced in *p*-xylene at different temperatures for 6 h. (**a**) 40 °C, (**b**) 50 °C, and (**c**) 60 °C (both PE21Br and RGO concentrations are 0.05 wt %).

Table 3. The average size of PE21M lamella formed on RGO based on the TEM images of 200 lamellae.

Sample (Crystallization Temperature)	Size of Lamellae (nm)	Thickness of Lamellae (nm)	Density (number/0.01 μm^2)
PE21M/RGO (40 °C)	5.0 ± 0.2	3.0 ± 0.3	7.0
PE21M/RGO (50 °C)	18.0 ± 1.2	5.3 ± 0.4	8.0
PE21M/RGO (60 °C)	16.0 ± 1.0	5.5 ± 0.3	5.0

3.1.3. PE21M/Nanofiller Composites Prepared with the Assistance of Supercritical CO_2

Supercritical CO_2 (SC CO_2) was used to follow the antisolvent effect on PE21M crystallization. PE21M/nanofiller nanocomposites were prepared in p-xylene at 50 °C with the assistance of SC CO_2 and the experimental pressure of SC CO_2 was tuned to 10, 15, and 20 MPa. The crystallization morphologies are shown in Figures 4 and 5. SWCNTs are apt to agglomerate and form bundles under pressure of SC CO_2, which gradually becomes serious as the SC CO_2 pressures rises (Figure 4c). Therefore, the diameter of the SWCNTs are much thicker than that produced without SC CO_2 (Figure 1). The average sizes of the PE21M lamellae formed on SWCNTs and RGO are listed in Tables 4 and 5. After 3 h isothermal crystallization with the assistance of SC CO_2, the diameter of PE21M crystals formed on SWCNTs at 50 °C and 10 MPa, 23.5 ± 1.0 nm, is larger than that of crystals via solution isothermal crystallization after 6 h (20.2 ± 1.1 nm, Table 1). As the SC CO_2 pressure increases, the diameter of PE21M formed on SWCNTs at 15 MPa increases to 28.3 ± 1.1 nm. Similarly, the lamellae size of PE21M formed on RGO at 10 MPa, 20 ± 1.3 nm, is also larger than that of crystals formed without the assistance of SC CO_2 after 6 h of isothermal crystallization (Table 2). This suggests that SC CO_2 can accelerate the lateral growth of lamellae. The diameter of the PE21M kebab decreases with the increasing pressure of SC CO_2, from 28.3 ± 1.1 at 15 MPa to 20.5 ± 1.1 nm at 20 MPa. This phenomenon has also been observed previously [36–38], which can be attributed to the amount and speed of the PE21M molecular precipitation being greatly increased and the crystal growth being restrained at the excessively high CO_2 pressure. The interval of PE21M kebabs formed on SWCNTs becomes smaller due to the larger deposition of the PE21M chain, from 18.9 ± 0.8 nm at 10 MPa and 20.6 ± 0.9 nm at 15 MPa to 16.5 ± 0.8 nm at 20 MPa. For the crystallization of PE21M on RGO, the average sizes of the rod-like crystals of PE21M/RGO nanocomposites are about 20.0 ± 1.3, 28.0 ± 1.3, and 30.0 ± 1.5 nm, respectively, for 10, 15, and 20 MPa, which increase with the increasing pressure of SC CO_2. The density of the rod-like crystals has the same variation rule, i.e., increases from 11/0.01 μm^2 at 10 MPa and 15/0.01 μm^2 at 15 MPa to 18/0.01 μm^2 at 20 MPa. This fact indicates that the large deposition of molecular chains under SC CO_2 is beneficial to the attachment of PE21M on the surface of RGO, which then promotes the crystal growth. We can also find from Figures 4 and 5 and Tables 4 and 5 that the thicknesses of the crystal lamellae produced with the assistance of SC CO_2 are almost consistent with those prepared by traditional solution crystallization. This suggests that SC CO_2 can promote the lateral growth of lamellae on CNTs and RGO, but it cannot change the thickness of the crystal lamellae, which is directly related to crystallization temperature. There is an optimum SC CO_2 pressure of 15 MPa, at which the rod-like crystal lamellae size is the maximum.

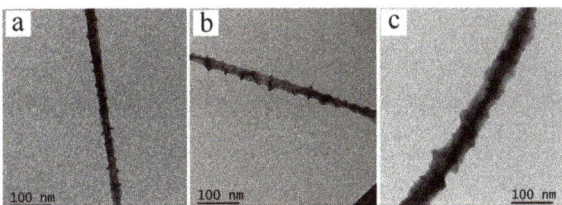

Figure 4. TEM images of PE21M/SWCNT nanocomposites produced at different SC CO_2 pressures: (**a**) 10 MPa, (**b**) 15 MPa, and (**c**) 20 MPa in p-xylene at 50 °C for 3 h.

Table 4. The average size of PE21M lamellae formed on SWCNT with the assistance of SC CO_2 based on the TEM images of 200 lamellae.

Sample (Pressure)	Diameter of Kebab (nm)	Thickness of Kebab (nm)	Interval of Kebab (nm)
PE21M/SWCNT (10 MPa)	23.5 ± 1.0	5.4 ± 0.6	18.9 ± 0.8
PE21M/SWCNT (15 MPa)	28.3 ± 1.1	5.5 ± 0.5	20.6 ± 0.9
PE21M/SWCNT (20 MPa)	20.5 ± 1.1	5.4 ± 0.7	16.5 ± 0.8

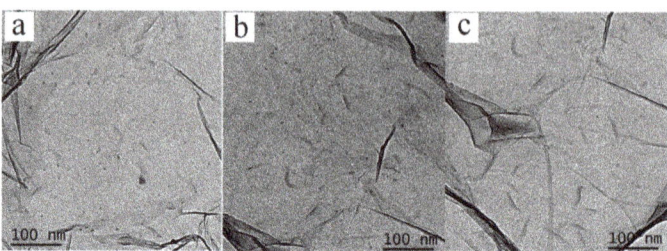

Figure 5. TEM images of PE21M/RGO nanocomposites produced at different SC CO_2 pressures: (**a**) 10 MPa, (**b**) 15 MPa, and (**c**) 20 MPa in *p*-xylene at 60 °C for 3 h.

Table 5. The average size of PE21M lamellae formed on RGO with assistance of SC CO_2 based on the TEM images of 200 lamellae.

Sample (Pressure)	Size of Lamellae (nm)	Thickness of Lamellae (nm)	Density (number/0.01 μm^2)
PE21M/RGO (10 MPa)	20.0 ± 1.3	5.4 ± 0.6	11
PE21M/RGO (15 MPa)	28.0 ± 1.3	5.6 ± 0.4	15
PE21M/RGO (20 MPa)	30.0 ± 1.5	5.6 ± 0.6	18

3.2. Thermal Behavior of PE21M/Nanofiller Composites

The melting behaviors of PE21M/nanofiller composites were measured by differential scanning calorimetry (DSC), as depicted in Figures 6 and 7. The melting peak temperatures of PE21M/nanofillers produced at 40 °C and 60 °C cannot be measured by DSC due to the very small size and quantity of epitaxial crystallized lamellae (Figures 1 and 3) and that of PE21M/nanofillers produced at 50 °C are 60.2 °C for PE21M/SWCNT and 60.5 °C for PE21M/RGO, which are all lower than that of PE21M (62.8 °C). This indicates that the thickness of lamellae formed on nanofillers is smaller than that of PE21M. We can also find from Table 6 that the melting temperature of PE21M/RGO nanocomposites is almost the same with that of PE21M/SWCNT under the same crystallization conditions, suggesting the analogical thickness of lamellae in both nanocomposites. The value of the heat of fusion (ΔH) of PE21M and nanocomposites are summarized in Table 6. The value of ΔH of PE21M/RGO composites is 0.16 J/g, much lower than 0.52 J/g of PE21M/CNT composites, which suggests that the quantity of crystals formed on RGO is much smaller than that formed on CNT.

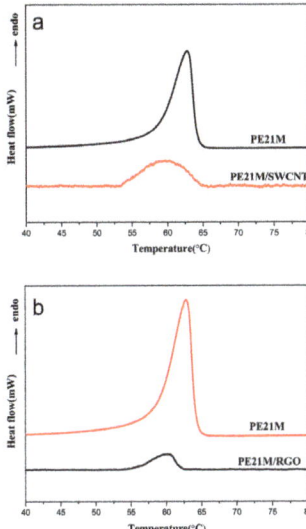

Figure 6. First heating curves of PE21M and PE21M/nanofiller composites in DSC measurement at a constant heating rate of 10 °C/min. (**a**) PE21M/SWCNT nanocomposites, and (**b**) PE21M/RGO nanocomposites.

Table 6. Melting data of PE21M and PE21M/nanofiller composites.

Sample	Tm (°C) Peak	Tm (°C) Onset	Onset-End (°C)	ΔH (J/g)
PE21M	62.8	58.0	6.8	18
PE21M/SWCNT	60.2	54.0	10.5	0.52
PE21M/RGO	60.5	56.2	5.7	0.16

PE21M/nanofiller composites produced at 50 °C and 15 MPa SC CO_2 pressure were taken as the example to investigate the influence of SC CO_2 on the melting behavior of PE21M crystals. We can find from Figure 6 and Table 7 that the melting peak temperature of PE21M/SWCNT nanocomposites produced at 50 °C is 60.5 °C for 15 MPa, which is almost consistent with that prepared by traditional solution crystallization (60.2 °C in Table 6). The same result is also observed in PE21M/RGO nanocomposites. The values of ΔH of PE21M/SWCNT and PE21M/RGO nanocomposites prepared at 50 °C and 15 MPa SC CO_2 pressure are 0.71 and 0.34 J/g, respectively, larger than that prepared by traditional solution crystallization (0.52 and 0.16 J/g shown in Table 6). This further suggests that supercritical CO_2 can promote the growth of lamellae on CNT and RGO, but the thickness of the crystal lamellae is directly related to crystallization temperature. The melting ranges of PE21M/SWCNT and PE21M/RGO nanocomposites prepared at 50 °C and 15 MPa SC CO_2 pressure are 13.5 and 12.2 °C, respectively, larger than that prepared by traditional solution crystallization (10.5 and 5.7 °C shown in Table 6). This is because the amount and speed of the PE21M precipitation from supercritical CO_2 is large, leading to the formation of more lamellae with inhomogeneous thickness.

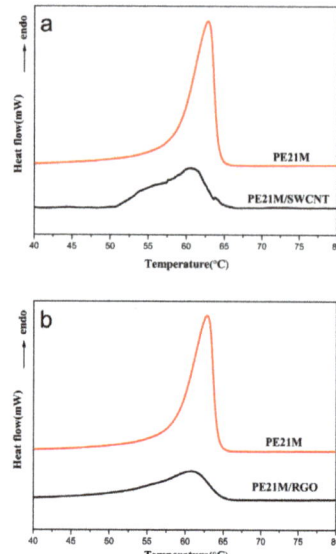

Figure 7. First heating curves of PE21M and PE21M/nanofiller composites produced at 50 °C and 15 MPa SC CO_2 pressures in DSC measurement at a constant heating rate of 10 °C/min. (**a**) PE21M/SWCNT nanocomposites, and (**b**) PE21M/RGO nanocomposites.

Table 7. Melting data of PE21M and PE21M/nanofiller composites produced at 15 MPa SC CO_2 pressure.

Sample	Tm (°C) Peak	Tm (°C) Onset	Onset-End (°C)	ΔH (J/g)
PE21M	62.8	58.0	6.8	18
PE21M/SWCNT	60.5	50.7	13.5	0.71
PE21M/RGO	60.7	52.5	12.2	0.34

3.3. Crystalline Structure of PE21M/Nanofiller Composites

X-ray diffractograms of PE21M/nanofiller composites are exhibited in Figure 8 together with the patterns of PE21M and nanofillers. As shown in Figure 8a, two peaks centered at 19.4° and 22.2° in the XRD pattern can be observed, assigned to pure PE21M, indexed as (100) and (010) reflections of the triclinic cell [31]. For pristine nanofillers, the peak at 26.8° is assigned to the (002) reflection of carbon. Two peaks of 19.4° and 22.2° in PE21M/nanofiller composites are weak compared to virgin PE21M. This can be due to the fact that the quantity of epitaxial crystallized lamellae is small (Figures 1 and 3), which results in the weak diffracted intensities of (100) and (010) reflections. All the peaks for PE21M and nanofillers are found in the XRD pattern of PE21M/nanofiller composites without the obvious peak shift. This indicates that the nanofillers have quite a small influence on the crystal structure of PE21M. The XRD of PE21M/nanofiller composites prepared with the assistance of supercritical CO_2 (Supplementary Data Figures S2 and S3) again show the same results, further indicating that supercritical CO_2 cannot change the crystal structure of PE21M/nanofiller composites although it can effectively promote the growth of lamellae on CNT and RGO.

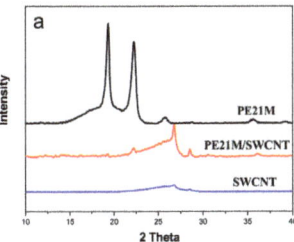 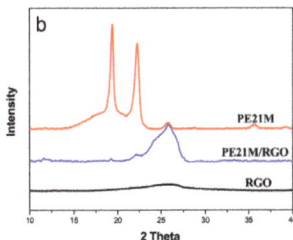

Figure 8. XRD diffractograms of PE21M, nanofillers, and PE21M/nanofiller composites. (**a**) PE21M/SWCNT nanocomposites, and (**b**) PE21M/RGO nanocomposites.

The substituent has a significant impact on the crystallization behavior of polyethylene since it acted as a defect of the chain to disturb the length of the continuous methylene sequences. This is confirmed by the fact that the orthorhombic lattice of PE21F is larger compared to that of HDPE, and the crystal lattice of PE21Cl and PE21Br are triclinic due to the accommodation of the bulky substituents [27,30]. Meanwhile, the melting temperatures of PE21F, PE21Cl, and PE21Br continuously decreased. With the substituent volume increasing in methyl (2 Å), the melting temperature of PE21M is much lower because of the lower-order triclinic form [27,30,31]. In our previous works [37,38], RGO could induce the structural transformation of PE21Cl and PE21Br from the triclinic form to the ordered orthorhombic form and generate extraordinary high melting temperatures due to the perfect lattice matching between polymers and RGO. SWCNT can only induce PE21Br to the orthorhombic form, and the melting temperatures of PE21Br/SWCNT nanocomposites are also lower than that of PE21Br/RGO nanocomposites because of the absent strict epitaxial crystallization. The results showed the capability that RGO induces substituted polymers to crystallize and generate an ordered chain packing structure because of the lattice matching. In this study, the triclinic lattice of PE21M cannot be changed by the induction of CNT and RGO. This suggests that the strong inductive effect from CNT and RGO cannot surmount the obstacle of methyl to the conformational adjustment of the PE21M molecular chain. Therefore, we conclude that polymers with the substituent volume greater than or equal to 2 Å (methyl) cannot be induced by CNT and graphene to form an ordered lattice structure. On the other hand, CNT and RGO show the different abilities that induce PE21M to nucleate, which can be explained by the different inducing mechanisms of both nanofillers. CNT induces the polymer to crystallize based on "soft epitaxy", that is, regardless of the lattice matching between the polymer chain and the graphitic sheet. Therefore, the number of the PE21M crystal nucleus formed on CNT was unaffected by the bulkier substituent, while, for two-dimensional RGO, the mechanism of RGO inducing the polymer is lattice matching, which will require adjusting the chain conformation to the surface of RGO. However, the bulky volume of methyl impedes the conformation adjustment of the PE21M chain, impacting the attachment of PE21M chain to the surface of RGO and nucleation. Consequently, the density of PE21M rod-like crystals formed on RGO is much smaller than that of kebab crystals on CNT.

4. Conclusions

The epitaxial crystallization behavior of PE21M on two different types of structural nanofillers, SWCNT and RGO, was investigated. The size and quantity of epitaxial crystallized lamellae formed on the surface of the nanofillers is small due to the bulkier substituent. The most suitable crystallization temperature for PE21M nanocomposites in p-xylene is 50 °C, at which the largest size and the most uniform lamellae of PE21M are formed. SC CO_2 can accelerate the growth of lamellae and improve the production of lamellae formed on nanofillers and the optimum SC CO_2 pressure for forming the maximum lamellae size is 15 MPa. The disparity in the density of crystals on CNT and RGO can be attributed to the different inducing mechanisms that show the different abilities of both nanofillers to induce PE21M to nucleate. The triclinic crystal lattice structure and crystallizable sequence length of PE21M remain unchanged in the PE21M/nanofillers composite system. Therefore, the methyl is the threshold volume of the substituent that could be induced by nanofillers to form an ordered lattice structure. This study helps us infer whether substituted or branched polymers can be induced by carbonaceous nanofillers to crystallize and generate an ordered chain packing structure according to the volume of the substituent or branch. Meanwhile, it offers a referral to fabricate polymer-carbonaceous nanocomposites with a controllable chain packing structure and expected physical properties.

Supplementary Materials: The following are available online at http://www.mdpi.com/2073-4352/8/4/168/s1; **Figure S1.** ^{1}H NMR for (**A**) PE21M, (**B**) unsaturated polymer, and (**C**) monomer; **Figure S2.** XRD diffractograms of PE21M and PE21M/SWCNT composite produced at 50 °C and 15 MPa SC CO_2 pressure; **Figure S3.** XRD diffractograms of PE21M and PE21M/RGO composite produced at 50 °C and 15 MPa SC CO_2 pressure.

Acknowledgments: This work is financially supported by the National Natural Science Foundation of China (U1532114), the China Postdoctoral Science Foundation (2017M621892), the Natural Science Foundation of Zhejiang Province (LY15B040003), the Natural Science Foundation of Ningbo Municipal (2015A610021), and the K.C. Wong Magna Fund at Ningbo University.

Author Contributions: This paper was accomplished based on the collaborative work of the authors. Weijun Miao performed the experiments, analyzed the data, interpreted the experimental results, and wrote the paper. Yiguo Li, Libin Jiang, Feng Wu, and Hao Zhu contributed to the experimental design. Hongbing Chen and Zongbao Wang supervised the entire research progress and analyzed the experimental results.

Conflicts of Interest: The authors declare no conflict of interest.

References

1. Wittmann, J.C.; Lotz, B. Epitaxial crystallization of polymers on organic and polymeric substrates. *Prog. Polym. Sci.* **1990**, *15*, 909–948. [CrossRef]
2. Chang, H.; Zhang, J.; Li, L.; Wang, Z.; Yang, C.; Takahashi, I.; Ozaki, Y.; Yan, S. A Study on the Epitaxial Ordering Process of the Polycaprolactone on the Highly Oriented Polyethylene Substrate. *Macromolecules* **2010**, *43*, 362–366. [CrossRef]
3. Wellinghoff, S.; Rybnikar, F.; Baer, E. Epitaxial crystallization of polyethylene. *J. Macromol. Sci. Phys.* **1974**, *10*, 1–39. [CrossRef]
4. Furuheim, K.M.; Axelson, D.E.; Antonsen, H.W.; Helle, T. Phase structural analyses of polyethylene extrusion coatings on high-density papers. I. Monoclinic crystallinity. *J. Appl. Polym. Sci.* **2010**, *91*, 218–225. [CrossRef]
5. Mittal, G.; Rhee, K.Y.; Park, S.J. The Effects of Cryomilling CNTs on the Thermal and Electrical Properties of CNT/PMMA Composites. *Polymers* **2016**, *8*, 169. [CrossRef]
6. Flores, A.; Poeppel, A.; Riekel, C.; Schulte, K. Evidence of a transcrystalline interphase in fiber PE homocomposites as revealed by microdiffraction experiments using synchrotron radiation. *J. Macromol. Sci. Phys.* **2001**, *40*, 749–761. [CrossRef]
7. Xu, J.Z.; Zhong, G.J.; Hsiao, B.S.; Fu, Q.; Li, Z.M. Low-dimensional carbonaceous nanofiller induced polymer crystallization. *Prog. Polym. Sci.* **2014**, *39*, 555–593. [CrossRef]
8. Yang, S.; Meng, D.; Sun, J.; Hou, W.; Ding, Y.; Jiang, S.; Huang, Y.; Geng, J. Enhanced electrochemical response for mercury ion detection based on poly(3-hexylthiophene) hybridized with multi-walled carbon nanotubes. *RSC Adv.* **2014**, *4*, 25051–25056. [CrossRef]
9. Xu, L.; Jiang, S.; Li, B.; Hou, W.; Li, G.; Memon, M.A.; Huang, Y.; Geng, J. Graphene Oxide: A Versatile Agent for Polyimide Foams with Improved Foaming Capability and Enhanced Flexibility. *Chem. Mater.* **2015**, *27*, 4358–4367. [CrossRef]
10. Zhu, W.; Yang, O.; Sun, J.; Memon, J.; Wang, C.; Geng, J.; Huang, Y. Scalable preparation of three-dimensional porous structures of reduced graphene oxide/cellulose composites and their application in supercapacitors. *Carbon* **2013**, *62*, 501–509.
11. Ning, N.; Fu, S.; Zhang, W.; Chen, F.; Wang, K.; Deng, H.; Zhang, Q.; Fu, Q. Metal-catalyst-free growth of carbon nanotubes/carbon nanofibers on carbon blacks using chemical vapor deposition. *Prog. Polym. Sci.* **2012**, *37*, 1425–1455. [CrossRef]

12. Mai, F.; Wang, K.; Yao, M.; Deng, H.; Chen, F.; Fu, Q. Superior Reinforcement in Melt-Spun Polyethylene/Multiwalled Carbon Nanotube Fiber through Formation of a Shish-Kebab Structure. *J. Phys. Chem. B* **2010**, *114*, 10693–10702. [CrossRef] [PubMed]
13. Li, A.; Zhang, C.; Zhang, Y.F. Thermal Conductivity of Graphene-Polymer Composites: Mechanisms, Properties, and Applications. *Polymers* **2017**, *9*, 437–454.
14. Deng, H.; Skipa, T.; Zhang, R.; Lellinger, D.; Bilotti, E.; Alig, I.; Peijs, T. Effect of melting and crystallization on the conductive network in conductive polymer composites. *Polymer* **2009**, *50*, 3747–3754. [CrossRef]
15. Deng, H.; Skipa, T.; Bilotti, E.; Zhang, R.; Lellinger, D.; Mezzo, L.; Fu, Q.; Alig, I.; Peijs, T. Preparation of High-Performance Conductive Polymer Fibers through Morphological Control of Networks Formed by Nanofillers. *Adv. Funct. Mater.* **2010**, *20*, 1424–1432. [CrossRef]
16. Kodjie, S.L.; Li, L.; Li, B.; Cai, W.; Li, C.Y.; Keating, M. Morphology and Crystallization Behavior of HDPE/CNT Nanocomposite. *J. Macromol. Sci. Phys.* **2006**, *45*, 231–245. [CrossRef]
17. Li, C.Y.; Li, L.; Cai, W.; Kodjie, S.L.; Tenneti, K.K. Nano hybrid shish-kebab: Periodically functionalize carbon nanotubes. *Adv. Mater.* **2005**, *17*, 1198–1202. [CrossRef]
18. Li, L.; Li, C.Y.; Ni, C. Polymer Crystallization-Driven, Periodic Patterning on Carbon Nanotubes. *J. Am. Chem. Soc.* **2006**, *128*, 1692–1699. [CrossRef] [PubMed]
19. Tracz, A.; Jeszka, J.K.; Kucinska, I.; Chapel, J.P.; Boiteux, G.; Kryszewski, M. Influence of the crystallization conditions on the morphology of the contact layer of polyethylene crystallized on graphite: Atomic force microscopy studies. *J. Appl. Polym. Sci.* **2002**, *86*, 1329–1336. [CrossRef]
20. Tracz, A.; Kucinska, I.; Jeszka, J.K. Formation of Highly Ordered, Unusually Broad Polyethylene Lamellae in Contact with Atomically Flat Solid Surfaces. *Macromolecules* **2003**, *36*, 10130–10132. [CrossRef]
21. Tracz, A.; Kucinska, I.; Jeszka, J.K. Unusual crystallization of polyethylene at melt/atomically flat interface: Lamellar thickening growth under normal pressure. *Polymer* **2006**, *47*, 7251–7258. [CrossRef]
22. Takenaka, Y.; Miyaji, H.; Hoshino, A.; Tracz, A.; Jeszka, J.K.; Kucinska, I. Interface Structure of Epitaxial Polyethylene Crystal Grown on HOPG and MoS$_2$ Substrates. *Macromolecules* **2004**, *37*, 9667–9669. [CrossRef]
23. Vadlamudi, M.; Alamo, R.G.; Fiscus, D.M.; Varma-Nair, M.J. Inter and intra-molecular branching distribution of tailored LLDPEs inferred by melting and crystallization behavior of narrow fractions. *J. Therm. Anal. Calorim.* **2009**, *96*, 697–704. [CrossRef]
24. Vadlamudi, M.; Subramanian, G.; Shanbhag, S.; Alamo, R.G.; Fiscus, D.M.; Brown, G.M.; Lu, C.; Ruff, C.J. Molecular Weight and Branching Distribution of a High Performance Metallocene Ethylene 1-Hexene Copolymer Film-Grade Resin. *Macromol. Symp.* **2009**, *282*, 1–13. [CrossRef]
25. Smith, J.A.; Brzezinska, K.R.; Valenti, D.J.; Wagener, K.B. Precisely Controlled Methyl Branching in Polyethylene via Acyclic Diene Metathesis (ADMET) Polymerization. *Macromolecules* **2000**, *33*, 3781–3794. [CrossRef]
26. Sworen, J.C.; Smith, J.A.; Berg, J.M.; Wagener, K.B. Modeling Branched Polyethylene: Copolymers Possessing Precisely Placed Ethyl Branches. *J. Am. Chem. Soc.* **2004**, *126*, 11238–11246. [CrossRef] [PubMed]
27. Boz, E.; Wagener, K.B.; Ghosal, A.; Fu, R.; Alamo, R.G. Synthesis and Crystallization of Precision ADMET Polyolefins Containing Halogens. *Macromolecules* **2006**, *39*, 4437–4447. [CrossRef]
28. Sworen, J.C.; Wagener, K.B. Linear Low-Density Polyethylene Containing Precisely Placed Hexyl Branches. *Macromolecules* **2007**, *40*, 4414–4423. [CrossRef]
29. Boz, E.; Ghiviriga, I.; Nemeth, A.J.; Jeon, K.; Alamo, R.G.; Wagener, K.B. Random, Defect-Free Ethylene/Vinyl Halide Model Copolymers via Condensation Polymerization. *Macromolecules* **2008**, *41*, 25–30. [CrossRef]
30. Boz, E.; Nemeth, A.J.; Wagener, K.B.; Jeon, K.; Smith, R.; Nazirov, F.; Bockstaller, M.R.; Alamo, R.G. Well-Defined Precision Ethylene/Vinyl Fluoride Polymers: Synthesis and Crystalline Properties. *Macromolecules* **2008**, *41*, 1647–1653. [CrossRef]
31. Rojas, G.; Inci, B.; Wei, Y.; Wagener, K.B. Precision polyethylene: Changes in morphology as a function of alkyl branch size. *J. Am. Chem. Soc.* **2009**, *131*, 17376–17386. [CrossRef] [PubMed]
32. Inci, B.; Lieberwirth, I.; Steffen, W.; Mezger, M.; Graf, R.; Landfester, K.; Wagener, K.B. Decreasing the Alkyl Branch Frequency in Precision Polyethylene: Effect of Alkyl Branch Size on Nanoscale Morphology. *Macromolecules* **2012**, *45*, 3367–3376. [CrossRef]
33. Gaines, T.W.; Nakano, T.; Chujo, Y.; Trigg, E.B.; Winey, K.I.; Wagener, K.B. Precise Sulfite Functionalization of Polyolefins via ADMET Polymerization. *ACS Macro Lett.* **2015**, *4*, 624–627. [CrossRef]

34. Tasaki, M.; Yamamoto, H.; Hanesaka, M.; Tashiro, K.; Boz, E.; Wagener, K.B.; Ruiz-Orta, C.; Alamo, R.G. Polymorphism and Phase Transitions of Precisely Halogen-Substituted Polyethylene. (1) Crystal Structures of Various Crystalline Modifications of Bromine-Substituted Polyethylene on Every 21st Backbone Carbon. *Macromolecules* **2014**, *47*, 4738–4749. [CrossRef]
35. Kaner, P.; Ruiz-Orta, C.; Boz, E.; Wagener, K.B.; Tasaki, M.; Tashiro, K.; Alamo, R.G. Kinetic Control of Chlorine Packing in Crystals of a Precisely Substituted Polyethylene. Toward Advanced Polyolefin Materials. *Macromolecules* **2014**, *47*, 236–245. [CrossRef]
36. Miao, W.; Lv, Y.; Zheng, W.; Wang, Z.; Chen, Z.R. Epitaxial crystallization of precisely fluorine substituted polyethylene induced by carbon nanotube and reduced graphene oxide. *Polymer* **2016**, *83*, 205–213. [CrossRef]
37. Miao, W.; Wang, Z.; Li, Z.; Zheng, W.; Chen, Z.R. Epitaxial crystallization of precisely chlorine-substituted polyethylene induced by carbon nanotube and graphene. *Polymer* **2016**, *94*, 53–61. [CrossRef]
38. Miao, W.; Wang, B.; Li, Y.; Zheng, W.; Chen, H.; Zhang, L.; Wang, Z. Epitaxial crystallization of precisely brominesubstituted polyethylene induced by carbon nanotubes and graphene. *RSC Adv.* **2017**, *7*, 17640–17649. [CrossRef]
39. Wang, B.; Li, H.; Li, Z.; Chen, P.; Wang, Z.; Gu, Q. Electrostatic adsorption method for preparing electrically conducting ultrahigh molecular weight polyethylene/graphene nanosheets composites with a segregated network. *Compos. Sci. Technol.* **2013**, *89*, 180–185. [CrossRef]
40. Zhang, Z.; Xu, Q.; Chen, Z.; Yue, Z. Nanohybrid Shish-Kebabs: Supercritical CO_2-Induced PE Epitaxy on Carbon Nanotubes. *Macromolecules* **2008**, *41*, 2868–2873. [CrossRef]

 © 2018 by the authors. Licensee MDPI, Basel, Switzerland. This article is an open access article distributed under the terms and conditions of the Creative Commons Attribution (CC BY) license (http://creativecommons.org/licenses/by/4.0/).

Review

One-Dimensional Zinc Oxide Nanomaterials for Application in High-Performance Advanced Optoelectronic Devices

Meng Ding [1,2], Zhen Guo [2,*], Lianqun Zhou [2,*], Xuan Fang [3], Lili Zhang [4], Leyong Zeng [5], Lina Xie [3] and Hongbin Zhao [6,*]

1. School of Physics and Technology, University of Jinan, 336 Nanxinzhuang West Road, Jinan 250022, China; dingmeng0207@163.com
2. CAS Key Lab of Bio-Medical Diagnostics, Suzhou Institute of Biomedical Engineering and Technology, Chinese Academy of Sciences, Suzhou 215163, China
3. School of Science and Engineering, The Chinese University of Hong Kong, Shenzhen 518172, China; fangxuan110@hotmail.com (X.F.); linna.xie@madison-tech.com (L.X.)
4. Shanghai Synchrotron Radiation Facility, Shanghai Institute of Applied Physics, Chinese Academy of Sciences, 239 Zhangheng Rd., Pudong, Shanghai 201800, China; zhanglili@sinap.ac.cn
5. College of Chemistry & Environmental Science, Hebei University, Baoding 071002, China; zengly@hbu.edu.cn
6. State Key Laboratory of Advanced Materials for Smart Sensing, General Research Institute for Nonferrous Metals, Beijing 100088, China
* Correspondence: guozhen@sibet.ac.cn (Z.G.); zhoulq@sibet.ac.cn (L.Z.); zhaohongbin@grinm.com (H.Z.)

Received: 27 April 2018; Accepted: 13 May 2018; Published: 18 May 2018

Abstract: Unlike conventional bulk or film materials, one-dimensional (1D) semiconducting zinc oxide (ZnO) nanostructures exhibit excellent photoelectric properties including ultrahigh intrinsic photoelectric gain, multiple light confinement, and subwavelength size effects. Compared with polycrystalline thin films, nanowires usually have high phase purity, no grain boundaries, and long-distance order, making them attractive for carrier transport in advanced optoelectronic devices. The properties of one-dimensional nanowires—such as strong optical absorption, light emission, and photoconductive gain—could improve the performance of light-emitting diodes (LEDs), photodetectors, solar cells, nanogenerators, field-effect transistors, and sensors. For example, ZnO nanowires behave as carrier transport channels in photoelectric devices, decreasing the loss of the light-generated carrier. The performance of LEDs and photoelectric detectors based on nanowires can be improved compared with that of devices based on polycrystalline thin films. This article reviews the fabrication methods of 1D ZnO nanostructures—including chemical vapor deposition, hydrothermal reaction, and electrochemical deposition—and the influence of the growth parameters on the growth rate and morphology. Important applications of 1D ZnO nanostructures in optoelectronic devices are described. Several approaches to improve the performance of 1D ZnO-based devices, including surface passivation, localized surface plasmons, and the piezo-phototronic effect, are summarized.

Keywords: zinc oxide; one-dimensional nanostructure; light-emitting diode; photodetector; localized surface plasmon; piezo-phototronic effect

1. Introduction

The small size of one-dimensional (1D) nanomaterials leads to unique electrical, mechanical, chemical, and optical properties that are attractive for application in nanoscience and nanotechnology. In particular, 1D zinc oxide (ZnO) is a representative nanomaterial with excellent properties, such as ultrahigh intrinsic photoelectric gain and multiple array light confinement. Thus, ZnO nanomaterials have been well studied in recent years with the goal of constructing advanced optoelectronic devices with improved performance [1]. It is anticipated that 1D ZnO nanomaterials could be used as an important basic unit of nanostructured systems. Several types of 1D ZnO nanostructures have been reported, including nanowires [2–6], nanorods [7–9], nanobelts [10–13], and nanotubes [14–16]. Considering actual application requirements, 1D ZnO nanomaterials are usually considered an ideal medium to construct optoelectronic devices. The performance of optoelectronic devices could be improved using 1D ZnO nanomaterials because of their excellent properties, such as carrier transport with high mobility, efficient light confinement, and light emission. Thus, 1D ZnO nanowires are usually used in devices for applications such as light emission, photon detection, and biochemical sensing. In addition, nanowire-like structures could be considered as an ideal system for physical studies investigating the carrier transport, light confinement, and energy loss behaviors of 1D confined objects. Therefore, 1D ZnO nanowires are favorable not only for examining the basic phenomena of low-dimensional systems, but also for constructing new types of nanodevices that exhibit high performance.

As a member of the II-VI family, ZnO possesses properties such as a wide direct band gap of 3.37 eV, large exciton binding energy of 60 meV at room temperature, and suitable features for feasible production of nanomaterials. Potential applications of ZnO materials include optical waveguides [17], varistors [18], solid-state lighting [19], gas sensors [20], and transparent conductive films. Because of its wide band gap, ZnO is a promising candidate material for use in solid-state optoelectronics that emit in the blue or ultraviolet (UV) spectral range, including lasers. In addition, the large exciton binding energy of ZnO means that it can display efficient excitonic emission even at room temperature [21].

Various fabrication methods of 1D ZnO nanomaterials have been developed, such as vapor transport [5], hydrothermal reaction [8,11,16,22], electrodeposition [12,23], chemical vapor deposition (CVD) [24,25], molecular beam epitaxy [26], and pulsed laser deposition [27]. These methods can be used to obtain samples on substrates such as Si, quartz, and sapphire. Each method has its specific merits and inevitable weaknesses. The gas vapor approaches can fabricate high-quality single-crystalline ZnO nanostructures. However, these processes require high temperatures (about 450–1000 °C) and often have other limitations, including poor sample uniformity, narrow substrate choice, and low product yield. Solution processing is one of the most commonly used methods to prepare 1D ZnO nanomaterials because of its low cost, controllability, and compatibility with large-area manufacturing.

Compared with other 1D nanomaterials, ZnO has three prominent advantages: first, it is a semiconductor with a piezoelectric effect, which is the basis for electric mechanical coupling sensors and inverters; second, the biosecurity and biocompatibility of ZnO is relatively high, so it can be used in medicine; third, ZnO has the largest range of known nanostructures, including nanowires [2–6], nanorods [7–9], nanobelts [10–13], nanotubes [14–16], nanoplates [28–30], nanospheres [31], nanotrees [32], nanoleaves [33], and nanosails [34]. In addition, ZnO tetrapod [35], microrod/microwire [36,37], porous array [38], matrix [39], and 3D network nanostructure [20,40] have been obtained. At present, many researchers are focusing on constructing devices based on ZnO nanomaterials. New devices based on 1D ZnO nanomaterials are constantly being developed, such as room-temperature lasers, light-emitting diodes (LEDs), photodetectors, sensors, and transistors. These devices containing 1D ZnO nanomaterials exhibit better performance than that of devices with thin films and bulk materials. The excellent performance of new devices based on 1D ZnO nanomaterials reveal their great potential for applications in many fields.

Previously, there were some review articles about ZnO nanostructure, for example, Wang et al. reviewed various nanostructures of ZnO using solid-vapor phase technique, growth mechanisms, the application of nanobelts in nanosensors, field effect transistors, and nanoresonators [1]. Yi et al., reviewing the fabrication method and doping of ZnO nanorods, also discussed the properties related to ZnO nanorods, including luminescence, field emission, electron transport, as well as various applications [41]. Wang comprehensively reviewed various ZnO nanostructures fabricated by wet chemical methods, their doping and alloying, patterned growth, application in catalysis, surface modification, sensing, optoelectronic, and so on [42]. The review mainly focuses on the effect of growth conditions on 1D ZnO nanostructures and achievements of 1D ZnO-based photodetectors and LEDs, and, in particular, the promising approaches to improve the performance of LEDs and photodetectors.

In this review, we briefly introduce the properties of ZnO, review the growth methods of 1D ZnO nanomaterials, and describe the progress and current status of photoelectric devices (LEDs and photodetectors) based on 1D ZnO nanomaterials. Section 2 provides a brief introduction to the basic properties of 1D ZnO nanostructures. Section 3 covers the fabrication of 1D ZnO nanostructures, including CVD, hydrothermal reaction, and electrodeposition, and describes the influence of growth parameters on the growth rate and morphology of 1D ZnO nanostructures. Finally, we focus on LEDs and photodetectors based on 1D ZnO nanomaterials. The recent achievements of ZnO-based photodetectors and LEDs with different device structures and promising approaches to improve the performance of LEDs and photodetectors—including surface passivation, localized surface plasmons resonance (LSPR), and the piezo-phototronic effect—are discussed.

2. Basic Properties of ZnO

ZnO is II-VI compound semiconductor with direct wide band gap of 3.37 eV at room temperature. ZnO has three typical crystal structures: wurtzite, zinc blende, and rock salt. The thermodynamically stable phase of ZnO at ambient temperature and pressure is wurtzite. The zinc blende structure can be stabilized by growth on cubic substrates. The rock-salt structure can be fabricated at high pressure [21].

Wurtzite ZnO has a hexagonal structure with lattice parameters a and c and belongs to the $P6_3mc$ space group. The lattice parameter a mostly ranges from 0.32475 to 0.32501 nm, and c typically ranges from 0.520642 to 0.52075 nm [43]. The deviation of ZnO lattice constants from the ideal values is probably caused by the presence of point defects, such as zinc antisites, oxygen vacancies, and dislocations [21]. As illustrated in the schematic of the wurtzite ZnO structure shown in Figure 1, both Zn and O atoms arrange in a hexagonal close-packed (hcp) pattern and are stacked alternately along the c-axis. Furthermore, each kind of atom is surrounded by four atoms of the other kind, and the O^{2-} or Zn^{2+} of two adjacent layers form a tetrahedral structure. The resulting non-centrosymmetric structure of wurtzite ZnO can result in piezoelectricity and the thermoelectric effect. A ZnO single crystal exhibits very strict crystallographic polarity. The polar plane of such a crystal is the basal plane, which is composed of the Zn (0001) plane and O (000$\bar{1}$) plane, inducing a normal dipole moment, spontaneous polarization along the c-axis, and a distinct surface energy [1,21].

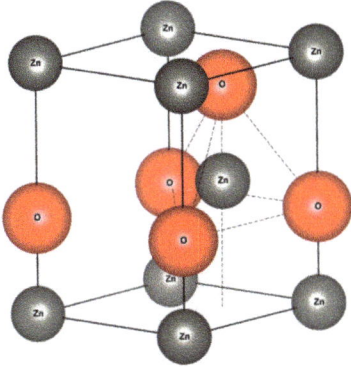

Figure 1. Schematic illustration of the wurtzite ZnO structure.

3. Growth of ZnO Nanostructures

ZnO probably possesses the richest variety of reported nanostructures, which are typically categorized into three groups: zero-dimensional, 1D, and two-dimensional. The four most common types of 1D ZnO nanostructures are nanowires [2–6], nanorods [7–9], nanobelts [10–13], and nanotubes [14–16]. Many approaches have been used to fabricate 1D ZnO nanostructures, such as CVD [4–6], hydrothermal processing [8,11,16,22], electrochemical deposition [12,23], and magnetron sputtering [44], with and without a catalyst.

3.1. Chemical Vapor Deposition

CVD is the one most commonly used methods to synthesize 1D ZnO nanostructures. In the CVD synthesis process, ZnO powder, Zn powder, or another Zn compound is used as the raw material after evaporation, reduction/oxidation, decomposition/combination, and other physical and chemical changes. The growth process is commonly carried out in a horizontal tube furnace, which consists of a horizontal tube heater, an alumina or quartz tube, a gas supply, and a control system. The fabrication of 1D ZnO nanostructures by CVD is generally affected by the synthesis parameters, such as the growth temperatures of the source and substrate, the distance between the source and substrate, heating rate, carrier gas (including gas species and flow rate), tube diameter, pressure, and atmosphere. Usually, the vapor–liquid–solid (VLS) process is used for 1D ZnO nanostructure growth via CVD, which requires a catalyst such as gold (Au) [5], copper (Cu) [45], stannum (Sn) [46], or cobalt (Co) [47]. The liquid droplet serves as a preferential site for absorption of the gas-phase reactant until the droplet is saturated and then becomes the nucleation site for crystallization. During growth, the catalyst droplet regulates the growth direction and controls the diameter of the nanowire [1]. Control of the nanowire diameter has also been achieved by varying the Au layer thickness. As shown in Figure 2, when the Au film coated on a substrate was 5 nm thick, the ZnO nanowires had diameters of 80–120 nm and lengths of 10–20 μm. When the Au film was 3 nm thick, the ZnO nanowires had diameters of 40–70 nm and lengths of 5–10 μm [5]. In addition to using external catalysts, Zn itself can act as a catalyst under certain conditions to prepare 1D ZnO nanostructures [48].

Figure 2. Scanning electron microscopy (SEM) images of ZnO nanowires grown on silicon substrates coated with gold films with a thickness of (**a**) ~5 nm and (**b**) 3 nm [5]. Reproduced with permission.

The influence of different catalyst thin films on the morphology and even nucleation process of ZnO nanomaterials varies. Osgoods et al. systematically studied ZnO nanowire growth using diverse metal catalysts (Au, Ag, Ni, and Fe) and substrates (Si, sapphire) with different structures and crystal orientation. SEM images of the nanowires are shown in Figure 3 [41–49]. The ZnO nanowires grew through the vapor–solid mechanism in the case of an iron (Fe) catalyst and via the VLS process using Au, Ag, or Ni as the catalyst. The ZnO nanostructures grown using different catalysts possessed differences in not only size and the ratio of length to diameter, but also the atomic composition ratio of Zn to O. Thus, the relative intensity of the oxygen vacancy-related emission in photoluminescence spectra of nanostructures grown on different thin films varied.

Figure 3. SEM images of ZnO nanowires fabricated on (**a**) Au-coated Si (111) and (**b**) Au-coated a-plane sapphire, (**c**) a-plane (110) sapphire coated with a 2.5 nm Ag film and (**d**) c-plane (001) sapphire coated with a 2.5 nm Ag thin film. SEM images (tilt angle of 60°) of ZnO nanowires grown on (**e**) an a-plane (110) sapphire substrate coated with a 2.5 nm Fe thin film and (**f**) a-plane sapphire with a 2.5 nm Ni thin film [49]. Reproduced with permission.

A ZnO thin film or other transition layer can be prepared on the substrate to facilitate the nucleation growth of 1D ZnO nanostructures. Fang et al. [6] demonstrated that ZnO thin films fabricated by electron-beam evaporation can provide nucleation sites to control the growth direction of ZnO nanowires. In addition, the density of ZnO nanowires could be adjusted by varying the thickness of the ZnO thin film, as shown in Figure 4. Yang and coworkers fabricated radial ZnO nanowire arrays on Si substrates with an amorphous carbon thin layer without a metal catalyst [50].

Figure 4. SEM images of ZnO nanorods grown on (**a**) a silicon wafer and (**b**) a ZnO nanostructured thin film [2]. Reproduced with permission.

3.2. Hydrothermal Method

The hydrothermal method is a very powerful technique for the growth of 1D ZnO nanomaterials, which involves chemical reaction of substances in solution at a certain temperature and pressure to fabricate the nanomaterials. This technique enables the synthesis of most materials with the required physical and chemical properties. Compared with other conventional preparation methods, the hydrothermal method has many advantages, such as low cost, simplicity, limited pollution, good nucleation and shape control, and low-temperature (<200 °C) growth at higher pressure in the presence of an appropriate solvent. In addition, the size distribution of materials obtained by the hydrothermal technique is uniform, with hardly any aggregation. The shape and size of the prepared materials are influenced by several processing parameters, such as the molar ratio and concentration of the reactant, temperature, and reaction time.

Vayssieres et al. [2,51] first used the hydrothermal method to fabricate ZnO nanorods in 2001. Equimolar mixed solutions of zinc nitrate ($Zn(NO_3)_2$) and hexamethylenetetramine ($C_6H_{12}N_4$, HMT) with different concentrations (1–10 mM) were heated at a constant temperature of 95 °C for several hours. The products were cleaned and then dried to give ZnO nanorod arrays. Various materials were used as substrates in their hydrothermal growth of ZnO nanostructures; for example, polycrystalline fluorine-doped tin oxide (FTO)-coated glass, single-crystalline sapphire, and Si/SiO_2 wafers with ZnO thin films. The experiments showed that the diameter of ZnO nanorods was controlled by the concentration of the precursors. As the concentration of the solution was lowered from 10 to 1 mM, the diameter of the nanorods decreased from 100–200 nm to 10–20 nm. SEM images of the ZnO nanorods fabricated by the hydrothermal method are shown in Figure 4. In this approach, only two kinds of chemical solutions with good solubility are needed. The two solutions can readily mix uniformly and promote ZnO formation, greatly simplifying and lowering the cost of the experimental process. The simple setup for hydrothermal growth of ZnO nanostructures is also beneficial to improve the purity of products and study the influence of synthetic parameters on nanostructure formation.

The method developed by Vayssieres and colleagues has since been greatly improved and remains in common use because of its simplicity. To improve the quality of the product and the controllability of the deposition process, a crystalline seed layer can be prepared on the substrate prior to hydrothermal growth. In 2003, Yang's group grew uniform ZnO nanowires on various substrates (e.g., Si (100) wafers and flexible polydimethylsiloxane films) using a two-step hydrothermal method [52]. The first step involved the preparation of a seed layer of ZnO nanocrystals with a diameter of 5–10 nm, which were spin-cast several times on the substrate to give a seed-layer thickness of 50–200 nm. The second step was ZnO nanowire growth by the hydrothermal method. The substrate with a seed crystal layer was placed in an open crystallizing dish filled with an equimolar solution of zinc nitrate hydrate and methenamine or diethylenetriamine (0.025 M) at 90 °C. Other research groups have prepared seed crystals using techniques such as pulsed laser deposition [14], magnetron sputtering [53], the sol–gel method [22,54], and atomic layer deposition [55]. Using small, homogeneous ZnO nanocrystals as a seed layer can provide nanowires that are small in diameter, well aligned, and in intimate contact with the substrate. In addition, the existing seeds could bypass the nucleation step, promoting growth on the seeds rather than nucleation in the homogeneous bulk solution. ZnO nanowires tend to grow wherever there are ZnO seeds; therefore, the density of nanowires is usually quite high [42]. As a result, the density of ZnO nanowires can be controlled by the thickness of the seed layer [22].

Although it is practical and effective to prepare ZnO nanostructures by the hydrothermal method using $Zn(NO_3)_2$ and HMT as raw materials, this approach does have shortcomings. For example, it is difficult to obtain ZnO nanomaterials with a large ratio of length to diameter. To increase the aspect ratio of the ZnO nanostructures, poly(ethylenimine) can be added to the reaction solution [56]. The addition of a low concentration of sodium citrate can alter the morphology of the ZnO nanorods. Sun et al. [57] prepared ZnO nanorods via cetyltrimethylammonium bromide (CTAB)-assisted hydrothermal oxidation of Zn metal at 180 °C. CTAB played important roles in the growth of ZnO nanorods, affecting the erosion process and growth orientation.

Kim et al. [8] prepared single-crystalline vertically aligned ZnO nanorod arrays on various substrates (Si, SiO$_2$, and FTO glass) using a polymer-templated hydrothermal method at a low temperature of 93 °C. To ensure that the ZnO nanorods grew along the c-axis [0001] direction and to minimize the crystal mismatch with the substrate, a c-oriented ZnO seed layer was first fabricated by the sol–gel method. The morphologies of the resulting nanostructures could be transformed from rodlike to pencil-like, needle-like, tubelike, treelike, and spherelike by exploiting the structural polarity of the ZnO nanorods and adjusting the growth parameters, including the seed layer, pH, and temperature, as summarized in Figure 5. Thus, Kim and coworkers developed a simple process to obtain ZnO nanorods with controllable size, density, and lattice.

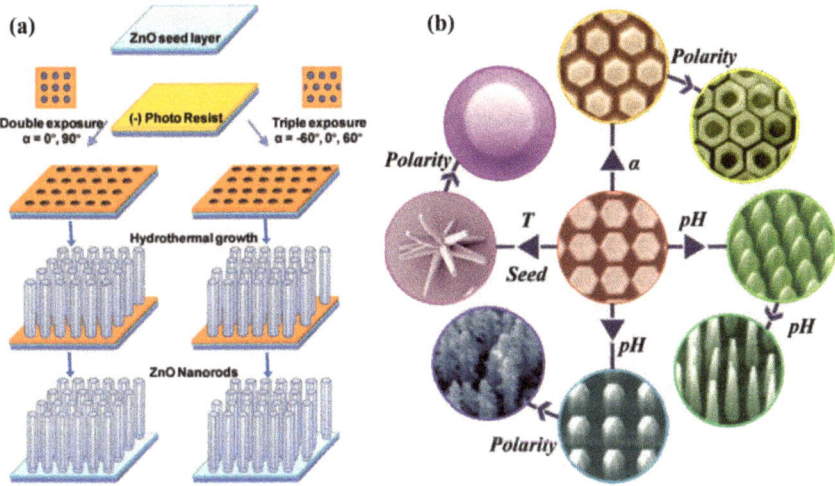

Figure 5. (a) Overview of the preparation process of periodic ZnO nanorods on various substrates (Si (100), SiO$_2$, and FTO glass); (b) high-resolution SEM images of the various ZnO nanostructures obtained using different reaction parameters [8]. Reproduced with permission.

3.3. Electrochemical Deposition

Electrochemical deposition has been widely used to obtain uniform and large-area growth of ZnO nanostructures [58]. Electrochemical deposition has many advantages, such as scalability, low cost, and easy handling [59]. Compared with ordinary chemical solution methods (e.g., hydrothermal methods), electrochemical deposition includes an external driving electric field in the reaction process, thus promoting reactions that are otherwise nonspontaneous. A standard three-electrode system with a saturated Ag/AgCl electrode as the reference electrode, Pt as the counter electrode (anode), and growth substrate as the working electrode is the typical setup for electrochemical deposition. In this case, the substrate serves as the cathode, so it must be conductive regardless of whether it is flat or curved. The substrate cathode is placed parallel to the anode in an aqueous solution of zinc salt (electrolyte). The chemical reaction is carried out in an electrolytic cell placed in a water bath to maintain a constant temperature.

Various groups have fabricated 1D ZnO nanostructures on FTO [60,61], indium-doped tin oxide (ITO) [62], polycrystalline Zn foil [63], and Si [64] substrates. Mari et al. [65] studied the influence of the bath temperature, current density, and deposition time on the growth rate of ZnO nanorods. Their results showed that the length of the nanorods increased with the current density and deposition time. Temperature had the opposite effect: the nanorods were longer when the deposition temperature was low. Xu and coworkers investigated the morphological control of ZnO nanostructures induced by addition of different capping agents, such as potassium chloride (KCl), ammonium fluoride (NH$_4$F), ammonium acetate (CH$_3$COONH$_4$), and ethylenediamine (EDA) [62]. The morphology of the ZnO structures evolved from hexagonal rods to rhombohedral rods and woven nanoneedles upon varying the compositions of the mixed capping agents, as depicted in Figure 6.

Figure 6. SEM images of the ZnO nanostructures formed by electrochemical deposition in solutions containing (**a**) 0.05 M Zn(NO$_3$)$_2$; (**b**) 0.05 M Zn(NO$_3$)$_2$ + 0.06 M KCl; (**c**) 0.05 M Zn(NO$_3$)$_2$ + 0.02 M NH$_4$F; (**d**) 0.05 M Zn(NO$_3$)$_2$ + 0.2 M NH$_4$F; (**e**) 0.05 M Zn(NO$_3$)$_2$ + 0.013 M ethylenediamine (EDA); and (**f**) 0.05 M Zn(NO$_3$)$_2$ + 0.06 M KCl + 0.01 M EDA [62]. Reproduced with permission.

Generally, zinc chloride (ZnCl$_2$) is used as the Zn^{2+} precursor when ZnO is deposited from the reduction of O$_2$ or H$_2$O$_2$. Zaera et al. [66] examined the effect of chloride (Cl$^-$) concentration, which was mainly supplied by the KCl supporting electrolyte, on the mechanism of ZnO nanowire growth driven by O$_2$ reduction. The rate of oxygen reduction decreased with rising Cl$^-$ concentration, and the deposition rate and dimensions of ZnO nanowire arrays could be tuned by varying the Cl$^-$ concentration of the solution, as shown in Figure 7. The effect of different anions (Cl$^-$, SO$_4^{2-}$, and CH$_3$COO$^-$) on the reduction of dissolved O$_2$ in the deposition solution has also been studied [67]. The observed differences in the O$_2$ reduction rate were explained by the different adsorption behaviors of the anions. Therefore, the different adsorption behaviors of the anions on ZnO surfaces could affect the generation rate of hydroxide ions and influence the morphology and growth rate of the ZnO nanowires. It was also demonstrated that the diameter and length of ZnO nanowires could be adjusted by varying the type of anions in the deposition solution. In particular, the nanowires grown in the presence of Cl$^-$ and CH$_3$COO$^-$ possessed low and high aspect ratios, respectively.

Figure 7. Cross-sectional SEM images of ZnO nanowire arrays obtained by electrochemical deposition using KCl concentrations of (**a**) 5×10^{-2} M; (**b**) 1 M; (**c**) 2 M; and (**d**) 3.4 M [66]. Reproduced with permission.

ZnO nanowires can also be fabricated by electrochemical deposition assisted by templates, including anodic aluminum oxide (AAO) [68–70], polycarbonate membranes [71,72], and porous films. AAO templates are widely used in the preparation of ZnO nanowires because of their simplicity and suitability for large-area, well-distributed nanowire fabrication. AAO templates can be prepared through a two-step anodization process [68,70]. The AAO template can be removed after nanowire growth, leaving the free-standing nanowires. Yi et al. [68] synthesized Zn nanowires consisting of the wurtzite hcp-Zn phase by electrodeposition on an AAO template with nanometer-sized pores. ZnO nanowires were then obtained by annealing the Zn nanowires at different temperatures for 10 h. Research has shown that this technique can provide not only ZnO nanowires, but also other semiconductor oxide nanostructures. However, ZnO and Al_2O_3 are amphoteric oxides, which makes the selective removal of the AAO membrane from the ZnO nanowires difficult. Polycarbonate membrane templates can be easily dissolved in chloroform, which makes them attractive to prepare free-standing 1D ZnO nanostructures. Wang and colleagues fabricated randomly distributed ZnO nanowires using a polycarbonate template [71]. Zeng et al. fabricated ZnO nanowire lines and bundle arrays on polystyrene (PS) template, as sown in Figure 8. The density of ZnO nanowires on Si substrate with 200 nm PS sphere template has decrease tendency with the heating time [73].

Figure 8. SEM images of 200 nm polystyrene (PS) sphere monolayer templates (**a**) and ZnO nanowires grown on 200 nm PS sphere template with beating time of 0 min (**b**); 3 min (**c**); 5 min (**d**), respectively [73]. Reproduced with permission.

4. Applications of 1D ZnO Nanostructures

4.1. Light-Emitting Diodes

An important feature of 1D ZnO nanomaterials is their excellent light transmission characteristics, which makes them attractive for use in high-efficiency UV LEDs and laser devices. ZnO nanomaterials display a direct wide band gap and large exciton binding energy at room temperature. UV LEDs have practical applications in solid-state lighting, UV photolithography, high-density data storage, and biomedical analysis [74]. The basis of LED construction is formation of a p–n junction. Unintentionally doped ZnO usually presents n-type conductivity. It is difficult to obtain p-type ZnO of high quality because of its low dopant solubility, deep acceptor level, and self-compensation effect. Therefore, the initial ZnO LEDs were constructed with other p-type semiconductors to form p–n heterojunctions that exhibited electroluminescence (EL). In the early stages, ZnO light-emitting devices were mainly based on thin films. After the thin-film LEDs were realized, many researchers shifted their attention to the study of LEDs containing 1D ZnO nanomaterials.

In 2006, Chang and coworkers fabricated a heterojunction device consisting of a ZnO nanowire array on an n-GaN film on a sapphire substrate and the p-type polymer poly(3,4-ethylenedioxythiophene)/poly(styrenesulfonate) (PEDOT/PSS). This hybrid LED exhibited multiple EL peaks at forward bias, including ZnO band edge emission at about 383 nm and other peaks centered at 430, 640, and 748 nm [75]. Zhang et al. [76] synthesized an n-type ZnO nanowire array on a p-type GaN film by CVD to produce a hybrid LED with high brightness. UV–blue EL emission was observed from the heterojunction diode, which displayed a blue shift as the forward bias increased. Tang et al. [77] fabricated an LED based on single n-ZnO/p-AlGaN heterojunction nanowires. A schematic illustration and SEM image of their LED are displayed in Figure 9a,b, respectively. When the injection current was 4 µA, UV EL emission with a peak at 394 nm was observed, as shown in Figure 9c.

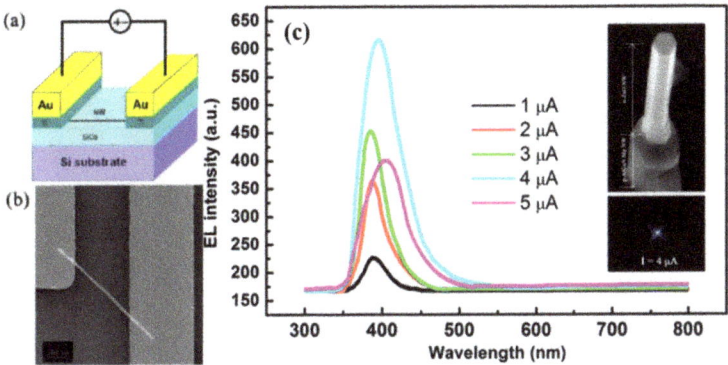

Figure 9. (a) Schematic illustration of the ZnO/AlGaN heterojunction light-emitting diode (LED); (b) SEM image of a single ZnO nanowire embedded in Au/Ti electrode films; (c) electroluminescence (EL) spectra of the LED at different injection currents. Insets: (upper) SEM image of a single nanowire; (lower) CCD image of the LED at an injection current of 4 µA [77]. Reproduced with permission.

Heterojunction diodes have been fabricated using n-type ZnO nanomaterials and p-type organic/inorganic materials such as GaN [78–80], AlGaN, Si [81,82], MgZnO [83], NiO [84], Cu$_2$O, SiC, N,N'-di(naphth-2-yl)-N,N'-diphenylbenzidine (NPB) [85], PEDOT/PSS [75], and [2-methoxy-5-(2-ethylhexyloxy)-1,4-phenylenevinylene] (MEH-PPV) [86]. Representative results for heterojunction LEDs containing 1D ZnO nanostructures, along with a brief description of the corresponding device structure, applied bias, and EL performance, are summarized in Table 1.

Table 1. 1D ZnO nanostructure-based heterojunction LEDs.

Device Structure	Turn-On Voltage	Threshold Voltage of Emission	Bias/Current	Wavelength of EL Peak	Ref.
ZnO nanowire arrays/P-GaN	3 V	4.4 V	5–7 V/0.5–1.5 mA	397 nm	[78]
ZnO nanowire arrays/P-GaN	3 V	4.4 V	6.5 V/2.6 mA	397 nm	[79]
n-GaN/ZnO nanorod/P-GaN	5 V		6 V	382 nm, 430 nm	[80]
Sb-doped p-ZnO nanowire arrays/n-GaN	3.74 V	4.0 V	16 V	391 nm	[87]
n-ZnO/p-AlGaN	2 V		4 µA	394 nm	[77]
ITO/n-ZnO nanorod array/p+-Si	5 V (5 mA)	6 V	9.5 V/29.5 mA	450 nm, 700 nm	[81]
n-ZnO nanorod/p-Si			9.5 V	387 nm, 535 nm	[88]
p-Zn$_{0.68}$Mg$_{0.32}$O:N/n-ZnO nanowire	4.2 V		5 mA/125 mA cm^2	390 nm	[83]
p-NiO/n-ZnO nanowire	2.5 V	7 V	15 V	385 nm, 570 nm	[84]
ZnO nanowire/polymer (PEDOT/PSS)		<3 V	3 V	383, 430, 640, 748 nm	[75]
ZnO nanorod/NPB			22 mA/cm^2	342 nm	[85]
ZnO nanorods/MEH-PPV			28 V	380, 580, 615, 640, 747 nm,	[86]

Many research groups are committed to obtaining p-type ZnO materials and have observed EL emission from ZnO p–n homojunctions. To date, p-type ZnO nanostructures have been realized by doping ZnO with group-V elements, including N [89], P [74,90], As [91,92], and Sb [93]. However, there have been relatively few reports of p–n homojunction LEDs based on 1D ZnO nanostructures.

Sun and coworkers obtained p-ZnO by As ion implantation into as-grown ZnO nanorods and then constructed ZnO nanorod homojunction LEDs [92]. A schematic illustration of the homojunction LED is presented in Figure 10a. At forward bias, UV emission originating from free-exciton recombination centered at about 380 nm was dominant, and a weak broad peak centered at about 630 nm was observed in EL spectra collected at room temperature. The second harmonic of UV emission centered at about 760 nm was also detected, as displayed in Figure 10b. The same group also used P as an acceptor dopant to fabricate ZnO nanorod homojunction LEDs by P ion implantation [74]. Strong UV emission attributed to near-band-edge emission and weak visible emission related to deep-level defects were observed from both devices at forward bias.

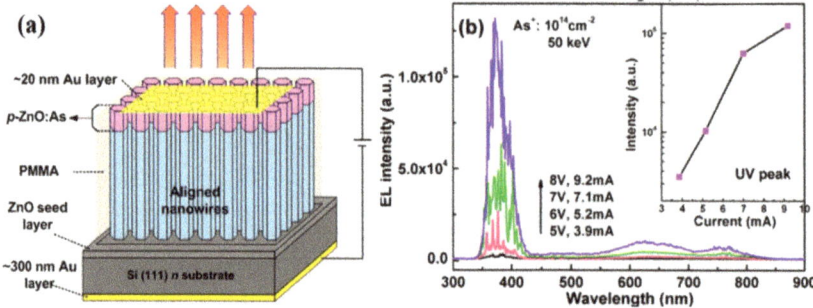

Figure 10. (a) Schematic illustration of an As-doped p-ZnO/n-ZnO nanorod homojunction LED; (b) EL spectra of a ZnO nanorod homojunction doped with As$^+$ at 50 keV and 10^{14} cm^{-2} [92]. Reproduced with permission.

However, there are some problems associated with heterojunction and homojunction LEDs based on 1D ZnO nanostructures, which can prevent the development and application of ZnO-based LEDs. For example, the heterostructured devices suffer from p-ZnO doping and large lattice mismatch. A Schottky-type LED based on Au/ZnO nanowires was constructed to overcome these issues [94]. A schematic illustration of the LED is shown in Figure 11a. The EL spectra in Figure 11b reveal that the LED displayed excitonic EL emission at about 380 nm at high forward bias.

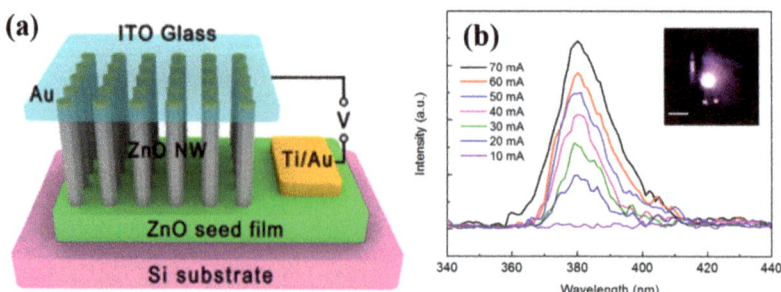

Figure 11. (a) Schematic illustration of the Au/ZnO nanowire Schottky LED; (b) EL spectra of the Schottky LED operating under different injection currents. Inset is a photograph of the device operating at an injection current of 70 mA [94]. Reproduced with permission.

Although there have been some encouraging achievements in LEDs based on 1D ZnO nanostructures, there are still some problems like the degradation of efficiency and stability caused by surface defects, surface adsorption, the low quality of p-type ZnO, and/or a high content of heterointerface defects. Surface passivation has been recognized as a promising approach to improve the EL performance of ZnO-based LEDs. In this approach, a dielectric/semiconductor material is coated on the surface of the 1D ZnO nanomaterial to form a core/shell nanostructure. Liu et al. [95]

fabricated two types of LEDs on p-GaN films using MgZnO-coated and bare ZnO nanorod arrays as active layers, as shown in Figure 12a. The EL performance of two types of LEDs is compared in Figure 12b. A strong emission peak at 387 nm was present in the EL spectra of both devices. The EL intensity of the MgZnO-coated LED was much higher than that of the uncoated one, which was attributed to the enhancement of the radiative recombination of the ZnO nanorods caused by the effective passivation of dangling bonds on the nanorod surface and the carrier confinement effect offered by the MgZnO coating. The stability of both devices exposed to ambient air for over a year at the same injection current of 5 mA was investigated at regular intervals; the results are displayed in Figure 12c. The ZnO near-band-edge UV emissions of both devices gradually decayed over time, but the decay rate of the uncoated device was faster than that of the MgZnO-coated LED. Therefore, the MgZnO-coated device exhibited higher stability than that of the uncoated one because surface adsorption on the nanorods was suppressed by the MgZnO coating. SiO_2 [96] and MgO [97] can also be used as passivation materials to improve the performance of 1D ZnO nanostructure LEDs.

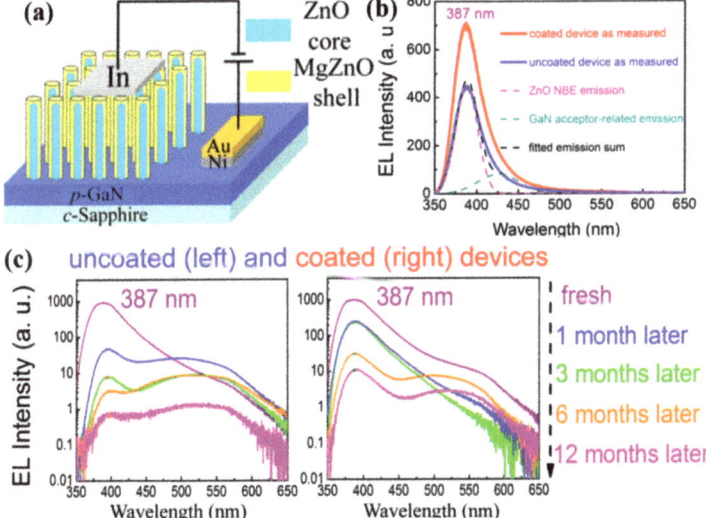

Figure 12. (a) Schematic illustration of an MgZnO-coated ZnO nanorod/p-GaN film LED; (b) EL spectra and Gaussian deconvolution analysis of the LEDs with and without an MgZnO coating under an injection current of 3.5 mA at room temperature; (c) EL spectra of LEDs with (right) and without (left) an MgZnO coating after exposure to air for different periods at an injection current of 5 mA [95]. Reproduced with permission.

Another effective approach to improve the luminous efficiency of LEDs is to introduce LSPs into the LED structure. Semiconductor excitons/photons can be resonantly coupled with the metal LSPs and then scattered/re-emitted into free space as radiation. This approach provides additional recombination/extraction pathways, which can increase the spontaneous radiation rate, internal quantum efficiency, and light extraction efficiency of the LEDs. Liu and coworkers fabricated ZnO nanorod array/p-GaN film heterostructure LEDs embedded in an Ag nanoparticle/polymethyl methacrylate (PMMA) composite [98]. A typical cross-sectional SEM image and schematic illustration of such an LED are shown in Figure 13a,b, respectively. The UV emission from ZnO excitons was enhanced more than 13-fold in EL spectra (Figure 13c). Lu et al. [99] constructed an Al nanoparticle-decorated n-ZnO nanorod/p-GaN film LED. They obtained 30-fold EL enhancement compared with that of the device without Al nanoparticles because of the effect of Al LSPs.

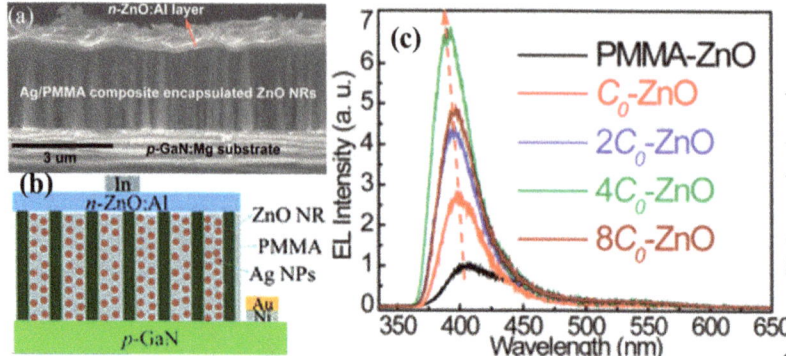

Figure 13. (a) Cross-sectional SEM image and (b) schematic illustration of a ZnO nanorod array/p-GaN film heterostructure LED embedded in a Ag nanoparticle/polymethyl methacrylate (PMMA) composite; (c) EL spectra of LEDs with different Ag nanoparticle concentrations in PMMA at an injection current of 5 mA at room temperature. $C_0 = 6.5 \times 10^{15}$ cm^{-3} [98]. Reproduced with permission.

To further improve the performance of LEDs, the two methods mentioned above can be combined. Zhang et al. [100] fabricated LSP-enhanced waveguide-type UV LEDs via sputtering Ag nanoparticles onto ZnO/MgZnO core/shell nanorod array/p-GaN film heterostructures. After decoration with Ag nanoparticles, the UV emission intensity of the ZnO/MgZnO core/shell nanorod array showed a ~9-fold enhancement compared with that of the device without Ag nanoparticles because of coupling with LSPs. The next year, the same group constructed LSP-enhanced UV LEDs by spin-coating Ag nanoparticles on ZnO/SiO$_2$ core/shell nanorod array/p-GaN heterostructures; a schematic diagram of this LED structure is illustrated in Figure 14a [96]. The EL intensity was enhanced when both Ag nanoparticles and a SiO$_2$ spacer layer were coated on the surface of the ZnO nanorods, as shown in Figure 14b. The effect of the SiO$_2$ spacer-layer thickness on the UV emission intensity was investigated. A maximum EL enhancement of ~7-fold was obtained when the thickness of the SiO$_2$ spacer layer was 12 nm. In contrast, a maximum PL enhancement of ~3.5-fold was achieved when the SiO$_2$ spacer layer was 16 nm thick. The EL enhancement was attributed to the combined effects of internal quantum efficiency improvement caused by exciton–LSP coupling and extraction efficiency enhancement induced by photon–LSP coupling.

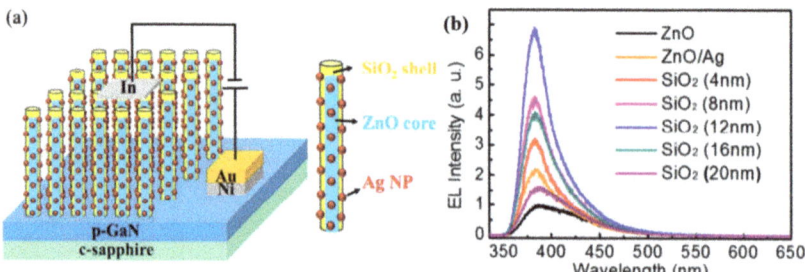

Figure 14. (a) Schematic diagram of a localized surface plasmon (LSP)-enhanced ZnO/SiO$_2$ core/shell nanorod array LED; (b) room-temperature EL spectra of bare and Ag-decorated ZnO core/shell nanorod arrays with different SiO$_2$ shell thicknesses under an injection current of 5 mA [96]. Reproduced with permission.

A near-UV LED based on a ZnO nanorod/MEH-PPV heterostructure using a ZnS layer as a buffer layer was reported by Wang and colleagues [101]. The intensity of near-UV EL was enhanced 10-fold compared with that of that of the device without a ZnS layer. In particular, the turn-on voltage was obviously decreased. In addition, the ZnS layer could induce recombination center of carriers, which regulated the EL wavelength of the heterojunction. Wei et al. [102] fabricated ZnO nanowire/p-GaN heterojunction LEDs containing ZnS particles to sensitize the ZnO nanowires. Localized states formed at the ZnO/ZnS interface, generating a built-in electric field, which captured electrons and holes at the ZnO/ZnS interface, where they recombined to result in EL emission from ZnO. Moreover, the piezo-phototronic effect can improve the efficiency of flexible ZnO nanowire/p-polymer hybridized LEDs [103,104]. Wang et al. [104] fabricated a hybrid LED based on a ZnO nanowire/p-polymer heterostructure. The external quantum efficiency of the hybrid LED at least doubled upon applying strain, reaching 5.92%.

4.2. Photodetectors

UV photodetectors are important photoelectric devices that occupy an indispensable position in today's information society. UV photodetectors are widely used in the military (e.g., UV control and guidance, UV alarms, and UV resistance), as well as in civil fields such as fire alarms, UV communication, ozone hole detection, and water purification. Therefore, the further development of UV detection technology has important implications in modern national defense and daily life. The importance and universality of UV photodetectors in military and civilian fields also make them a focus of current research.

ZnO has a wide direct band gap at room temperature of about 3.37 eV, along with high chemical and thermal stability, strong ability to resist radiation damage, and low content of electron-induced defects. Because of these favorable features, ZnO is widely used in UV detectors. Early ZnO UV detectors were mainly based on membrane structures [105]. However, with the development of low-dimensional materials, photodetectors based on 1D nanomaterials have attracted interest. Compared with UV detectors based on thin films and bulk materials, those based on 1D ZnO nanostructures display higher responsivity and internal photoconductive gain. This is because 1D nanomaterials have a large surface-to-volume ratio and the presence of deep-level surface trap states can prolong the lifetime of photogenerated electrons and holes. In addition, the low dimensionality of the active area can shorten the transit time of carriers [4,106]. There are several types of photodetector structures, such as photoconductive, metal–semiconductor–metal (MSM), Schottky [107], and p–n junction [108–110] photodetectors.

In 2002, Yang and coworkers first detected the UV response of an individual ZnO nanowire [111]. The conductance of a single ZnO nanowire was very sensitive to UV light. The resistivity of a ZnO nanowire reversibly changed by 4–6 orders of magnitude under UV illumination. In 2007, Soci et al. [4] reported a UV photodetector based on a single ZnO nanowire with an internal photoconductive gain as high as $\sim 10^8$. While the photoconductive detector showed slow response times because of possible persistent photoconductivity effects, it usually worked with an external power source.

To realize ZnO-based UV photodetectors with high sensitivity and fast response time, a Schottky contact instead of an ohmic contact was adopted [107,112–115]. For example, Zhou and coworkers fabricated a Schottky junction-type photodetector that used ZnO nanowire and Pt metal electrode to form Schottky contact [98]. The UV sensitivity of the photodetector was enhanced by $\sim 10^4$, and its recovery time was markedly shortened from ~417 to ~0.8 s, as shown in Figure 15.

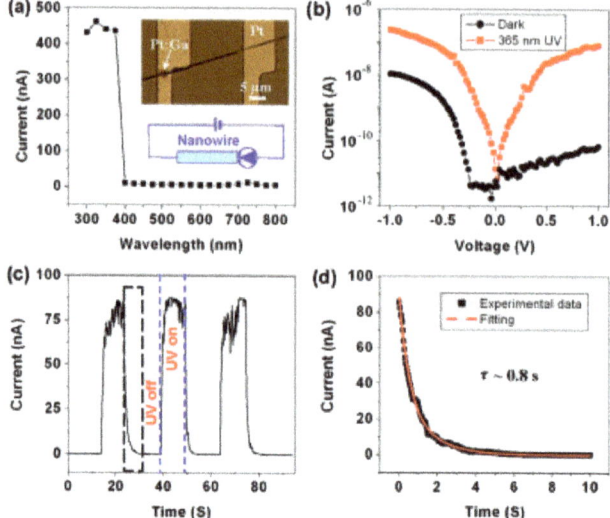

Figure 15. (a) The current of a ZnO nanowire UV detector as a function of the wavelength of incident light. Insets are an optical microscope image (upper) and schematic structure (lower) of the Schottky-type ZnO nanowire device; (b) current–voltage characteristics of the photodetector in the dark and under 365 nm UV illumination; (c) time dependence of the device photocurrent under on/off switching of 365 nm UV illumination (the applied bias was 1 V); (d) experimental data points and fitting curve of the photocurrent decay [113]. Reproduced with permission.

Nie and coworkers [115] reported a Schottky junction UV photodetector based on transparent monolayer graphene (MLG)/ZnO nanorods. The photodetector showed a ratio of photocurrent to dark current of approximately 3 orders of magnitude under 365 nm UV illumination at a bias of −1 V. The responsivity and photoconductive gain of the detector were 113 A W^{-1} and 385, respectively (Figure 16). The response speed of the UV photodetector was rather fast, with rise and fall times of 0.7 and 3.6 ms, respectively, which was attributed to the formation of a Schottky barrier, high crystal quality of the ZnO nanorod array, and strong light-trapping effect of the structure.

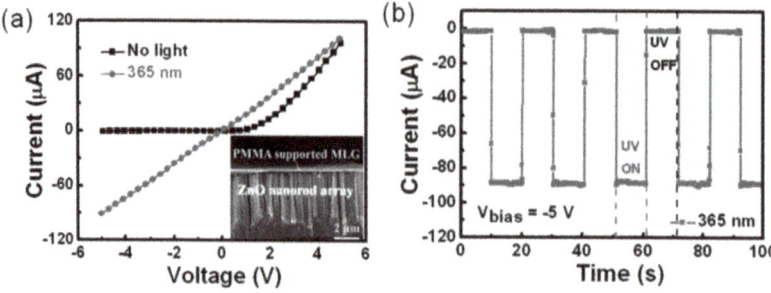

Figure 16. (a) Current–voltage characteristics of the monolayer graphene (MLG)/ZnO nanorod UV photodetector in the dark and under 365 nm UV illumination with an incident intensity of 100 μW cm^{-2}. Inset is a cross-sectional SEM image of the PMMA-supported MLG film/ZnO nanorod Schottky photodetector; (b) response times of the MLG/ZnO nanorod UV photodetector. The applied bias was −1 V [115]. Reproduced with permission.

Several kinds of 1D ZnO nanostructure-based p–n junction photodetectors have been reported [108,110,116,117], which possess the advantages of low applied fields and no oxygen dependency over Schottky junction detectors. In particular, p–n junction photodetectors work at rather high bias voltage to inhibit the recombination of electron–hole pairs, which increases the detection capability of UV signals [108,118]. Tahani et al. [108] developed a photodetector based on p-Si/n-ZnO nanotube heterojunctions, which is shown in Figure 17. Under 365 nm UV illumination with an incident power intensity of 1 mW cm^{-2} at a bias of -2 V, the responsivity and detectivity of the detector were 21.51 A W^{-1} and 1.26×10^{12} cm Hz$^{1/2}$ W^{-1}, along with an external quantum efficiency of 73.1×10^2% (effective area = ~0.79 mm^2). The rise and fall times of the heterojunction device were 0.44 and 0.599 s, respectively.

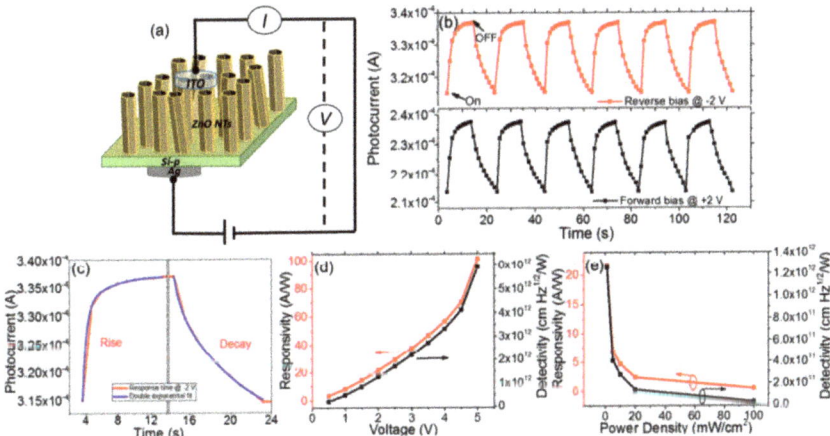

Figure 17. (a) Schematic diagram of the structure of a p-Si/n-ZnO nanotube UV photodetector; (b) detector photoresponse under 365 nm UV illumination with an incident intensity of 1 mW cm^{-2} at a bias of ± 2 V; (c) response times of the photodetector under 365 nm illumination (1 mW cm^{-2}) at a bias of -2 V; (d) device responsivity and detectivity as functions of applied bias under 365 nm illumination (1 mW cm^{-2}); (e) responsivity and detectivity as functions of light intensity under 365 nm illumination at a bias of -2 V [108]. Reproduced with permission.

Recently, self-powered ZnO photodetectors have attracted research interest because they can work in some poor conditions without an external power source. According to their interface features, self-powered photodetectors have three types of structures: Schottky junction, p–n junction, and photoelectrochemical. Compared with the other two types, ZnO-based p–n junction UV detectors are more suitable for development as self-powered photodetectors because of their much lower applied fields, faster response times, and high stability. Table 2 summarizes representative results for reported 1D ZnO nanostructure-based self-powered UV photodetectors, accompanied with a brief description of the corresponding device characteristics, detection wavelengths, power intensity, and photodetector performance.

Table 2. Recently reported self-powered 1D ZnO nanostructure-based UV photodetectors.

Device Structure	Light of Detection, Power	UV–Visible Rejection Ratio	On/Off Ratio	Responsivity	Rise/Decay Time	Ref.
ZnO/Sb-doped ZnO nanowire	365 nm, 0.3 mW cm^{-2}		26.7	26.5	30 ms/30 ms	[119]
n-ZnO nanowire/P-GaN film					20 μs/219 μs	[120]

Table 2. Cont.

Device Structure	Light of Detection, Power	UV–Visible Rejection Ratio	On/Off Ratio	Responsivity	Rise/Decay Time	Ref.
n-ZnO nanorods/i-MgO/p-GaN	350 nm	R350 nm/R500 nm) = 34.5	8000	0.32 A/W		[121]
ZnO nanorods/Si				UV/Visible: 0.3 A W^{-1}/0.5 A W^{-1}		[122]
ZnO Nanowire/CuSCN	370 nm, 100 mW/cm^2	R370 nm/R500 nm) = 100		0.02 A/W		[123]
ZnO Nanorod/CuSCN	380 nm	100		0.0075 A W^{-1}	5 ns/6.7 us	
PEDOT:PSS/ZnO micro-/nanowire			1000			[124]
n-ZnO/p-NiO core–shell nanowire				0.493 mA W^{-1}	1.38 ms/10.0 ms and 30.3 ms	[125]
p-NiO/n-ZnO nanorod array	355 nm			0.44 mA W^{-1}	0.23 s/0.21 s	[126]
p-NiO/n-ZnO nanowire	380 nm, 0.36 mW/cm^2			1.4 mA/W		[127]
$CH_3NH_3PbI_3$ perovskite/ZnO nanorod				3.9 A W^{-1} 300 nm	0.3 s/0.7 s	[128]
ZnO/Cu_2O nanowire						[129]

Various approaches have been adopted to improve the photoresponse performance of 1D ZnO nanostructure photodetectors for use in practical applications. Generally, surface and interface carrier transport modulation are the main routes used to raise photoresponsivity [130]. For instance, Lao et al. [131] manipulated and functionalized the polymer polystyrene sulfate (PSS) on the surface of ZnO nanobelts using a layer-by-layer self-assembly method. The sensitivity of the PSS-coated ZnO nanobelt-based detector was improved by close to 5 orders of magnitude compared with that of the device lacking PSS, which was ascribed to the energy levels introduced by PSS serving as a "hopping" state or bridge for electron transfer, effectively increasing the excitation probability of an electron to the conduction band.

In addition, functional devices consisting of a noble metal (e.g., Au [132,133] and Ag [134–137]) coated on semiconductor nanostructures have inspired extensive research effort. Such devices can realize superior optoelectronic properties to those without a metal coating because their LSPR effect leads to strong scattering and absorption of incident light, efficient separation of photogenerated electron–hole pairs, and rapid transport of charge carriers at the metal/semiconductor interface. Liu and coworkers fabricated ZnO nanowire photodetectors coated with Au nanoparticles [132]. The ratio of photocurrent to dark current of the ZnO photodetector with Au nanoparticles reached up to 5×10^6, which was much larger than that of the device without Au nanoparticles (around 10^3). Moreover, the decay time of the device decreased from ~300 s to ~10 s after coating with Au nanoparticles.

Al is an abundant and inexpensive metal that is a better plasmonic material than Au and Ag in the UV range because of the negative real part and relatively low imaginary part of its dielectric function [138]. Lu et al. [139] used LSP resonance mediated by Al nanoparticles to optimize the UV response of a ZnO nanorod array photodetector. The responsivity of the photodetector was improved from 0.12 to 1.59 A W^{-1} after modification with Al nanoparticles and the ratio of photocurrent to dark current was enhanced 6-fold compared with that of the bare one, as displayed in Figure 18. Furthermore, the rise and decay times of the photodetector decreased from 0.8 and 0.85 s, respectively, to less than 0.03 and 0.035 s, respectively, after Al nanoparticle decoration.

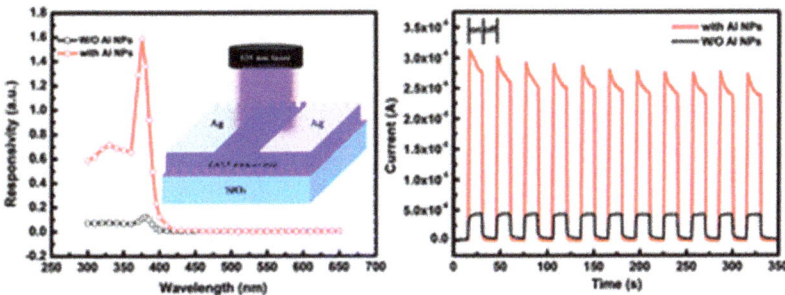

Figure 18. (**a**) Photoresponsivity spectra of ZnO nanorod photodetectors without and with Al nanoparticles. Inset is a schematic of the device structure; (**b**) time response of the devices with and without Al nanoparticles under 325 nm illumination at a bias of 5.0 V [139]. Reproduced with permission.

The transition metal Co is an effective candidate material for the surface modification of photodetectors. Buddha and coworkers fabricated a UV photodetector based on a Co-coated ZnO (Co–ZnO) nanorod array [140]; a schematic of the device is displayed in Figure 19a. The presence of defect states in the Co–ZnO nanorod detector raised the photocurrent compared with that of the device with bare ZnO nanorods. The Co–ZnO nanorod photodetector was sensitive to an external magnetic field; its response current increased by about 186% under an applied magnetic field of 2400 G compared with that without an applied magnetic field at the same bias voltage (Figure 19b). Furthermore, the response and recovery times of the photodetector decreased from 38 and 195 s, respectively, to less than 1.2 and 7.4 s, respectively, upon Co decoration, as shown in Figure 19c,d.

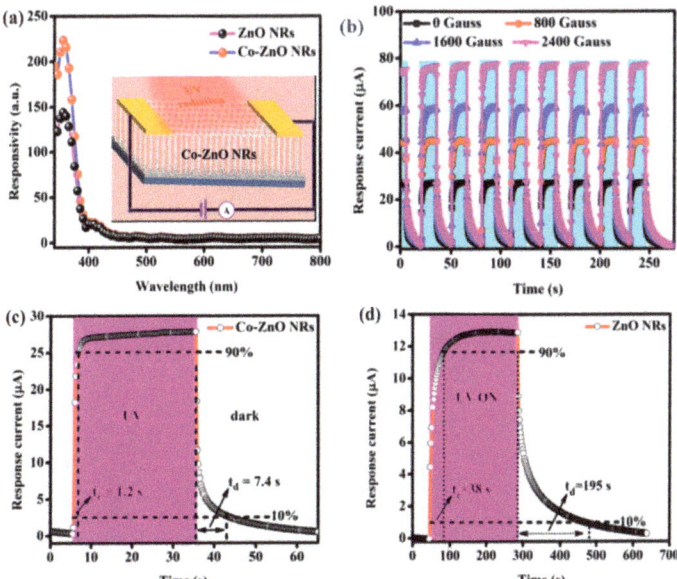

Figure 19. (**a**) Photoresponse spectra of photodetectors based on bare and Co-coated ZnO nanorods. The inset shows a schematic of the device; (**b**) cyclic response of the Co-coated ZnO nanorod photodetector with and without an external magnetic field at an applied bias of 5 V. The recovery times of (**c**) Co-coated and (**d**) bare ZnO nanorod photodetectors [140]. Reproduced with permission.

Recently, many researchers have attempted to improve the performance of photodetectors based on ZnO 1D nanostructures through the piezo-phototronic effect [141–146] and obtained some noticeable achievements. ZnO with a non-centrosymmetric crystal structure possesses high polarization. Therefore, the generation, transport, separation, and recombination process of carriers can be controlled by the piezopotential caused by the strain in this piezoelectric semiconductor material. It is worth noting that the coupled piezoelectric polarization, optoelectronic, and semiconducting properties of 1D ZnO nanomaterials offer an opportunity to construct functional devices with improved optoelectronic performance. Wang et al. [146] demonstrated that the responsivity of an MSM ZnO micro-/nanowire photodetector was enhanced through the piezoelectric effect. Zhang et al. [143] fabricated a self-powered photodetector based on a metal–insulator–semiconductor (MIS) Pt/Al$_2$O$_3$/ZnO structure; a schematic diagram of the device is shown in Figure 20a. The Schottky barrier height (SBH) was markedly enhanced to 0.739 eV by introducing the ultrathin dielectric layer (Al$_2$O$_3$) at the interface. The SBH could be actively adjusted by the modulation of the piezopolarization-induced built-in electric field under compressive strain, as shown in Figure 20b. The responsivity and detectivity of the photodetector increased from 0.644 µA W^{-1} and 2.92 × 10^6 cm Hz$^{0.5}$ W^{-1}, respectively, without compressive strain to 0.644 µA W^{-1} and 2.92 × 10^6 cm Hz$^{0.5}$ W^{-1}, respectively, under compressive strain of −1.0%.

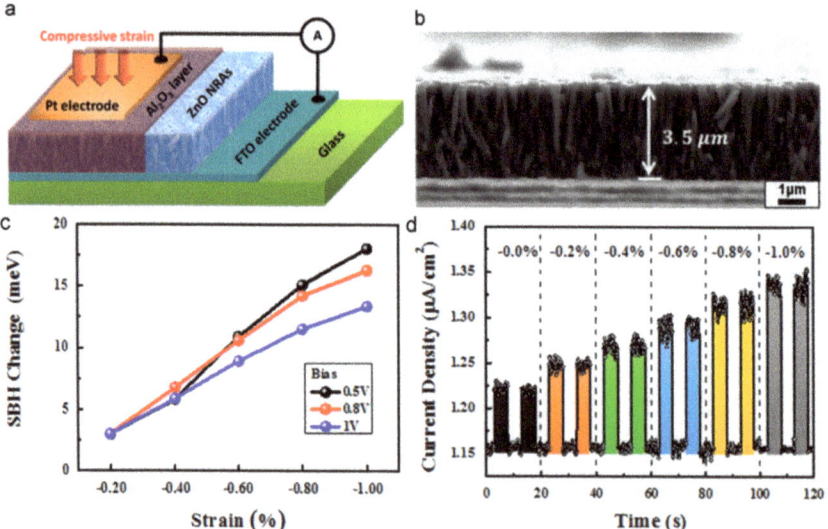

Figure 20. (**a**) Schematic diagram of the Pt/Al$_2$O$_3$/ZnO nanorod metal–insulator–semiconductor (MIS) photodetector; (**b**) typical cross-sectional SEM image of the ZnO nanorod array; (**c**) calculated Schottky barrier height (SBH) change under compressive strain at different biases of 0.5, 0.8, and 1 V; (**d**) device photoresponses under 100 mW cm^{-2} illumination and different compressive strains [143]. Reproduced with permission.

5. Conclusions

In conclusion, low-dimensional ZnO nanomaterials could be suitable for use in advanced optoelectronic devices, such as LEDs and photodetectors, because of their favorable properties. This review focused on the fabrication methods and growth parameters that influence the properties of the obtained 1D ZnO nanostructures. The fabricated nanomaterials could be used to construct novel devices with improved performance for certain applications. The properties, preparation, and device applications of 1D ZnO nanostructures were reviewed in detail. Importantly, several promising approaches to improve device performance, including surface passivation, LSPs, and piezo-phototronic effects, were discussed.

Considering the wide application potential of 1D ZnO nanomaterials and their excellent performance in devices, they can satisfy the requirements of "smaller, faster, colder" for electronics, which is expected to be extended to practical applications in the near future. The applications of 1D ZnO nanomaterials have expanded greatly in recent years, and the performance of various devices has been continuously improved. Novel high-performance devices based on 1D ZnO nanomaterials will certainly be an important research focus of the nanotechnology industry in the 21st century. With the improvement of preparation technology, deepening of research, constant improvement of device performance, and progress of nanotechnology industrialization, 1D ZnO nanomaterials will play an increasingly important role in the fields of energy, environmental protection, information science technology, biomedicine, security, and national defense.

The fundamental properties and applications of 1D ZnO nanomaterials have been explored extensively to optimize their advantages, providing a route to meet practical requirements in nanoscience and nanotechnology. Because 1D ZnO nanomaterials possess a large binding exciton energy, wide band gap, high carrier mobility, and non-centrosymmetric structure, they can be used to couple different properties such as semiconductivity, piezoelectricity, and photoexcitation together to develop next-generation devices with attractive properties such as intelligent features and high efficiency. At present, much attention has been paid to LEDs and photodetectors based on 1D ZnO nanomaterials, and more and more interdisciplinary research has been conducted, such as biochemical applications of 1D ZnO nanomaterials. It is anticipated that the carrier transport in 1D ZnO nanomaterials can be well tuned to maximize conversion efficiency. Through the combination of semiconducting properties and piezo-phototronic and size effects, the energy band structure of devices could be tuned to modulate charge carrier transport behavior, and device performance could be improved by controlling the SBH. Therefore, coupled effects can be readily applied to 1D ZnO nanomaterial-based devices to improve performance. We anticipate that research on 1D ZnO nanomaterial-based optoelectronic devices, such as LEDs and photodetectors, will yield further progress.

6. Perspective

There has been extensive and fundamental research performed on 1D ZnO nanomaterials and devices, resulting in methods to construct intelligent, smart, functional units with advanced performances for potential applications. One-dimensional ZnO nanomaterials generally have non-central symmetry crystal structure, high crystal quality, large aspect ratio, high carrier mobility, wide band gap, and other characteristics that can be applied for coupling with properties such as photoexcitation, piezoelectricity, and semiconductivity for designing next-generation, newly emerging devices. At present, many types of 1D ZnO nanodevices have emerged with promising applications, for example, energy harvesting, piezo-phototronic light emitting, photodetecting, nano-biosensing, and so on, based on 1D ZnO nanomaterials. When considering the design factors for building several kinds of nanodevices based on 1D ZnO nanomaterial, charge carrier generation, transport, and recombination dynamics are generally considered by studying band gap structure, non-central symmetry induced piezotronic effects, surface or interface contact properties, and so on. Thus, conversion efficiency could be key in the overall consideration for building advanced optoelectric devices with improved performance. Until now, it was anticipated that the charge carrier transport behaviors of the 1D ZnO nanomaterials could be well modulated for building devices based on piezotronic and piezo-phototronic effects. Self-powered 1D ZnO nanodevices could be constructed with outperforming properties based on the above effects, thus, next-generation, intelligent products with smart volume could be built by applying the mentioned ideas. By optimally combining their properties based on 1D ZnO nanomaterials, the performance of these devices, such as nanogenerators, LEDs, photodetectors, nanosensors, and memory devices, can be improved by modulating charge carrier transport behavior, interface barrier structures, piezotronic effect, and so on. It is believed that the coupled effects can be well applied for modulating carrier transport behaviors in 1D ZnO nanodevices for improving their performance. Further, newly emerging 1D ZnO nanodevices could be constructed for promising applications by studying its known effects and other properties through doping and surface modification. Finally, we anticipate that research work on 1D ZnO nanodevices—such as LEDs, photodetectors, solar cells, nanosensors, nanogenerators, and others—will yield continuous progress, and novel optoelectric, light emitting, electric nanodevices could be created through further efforts.

Author Contributions: The manuscript was written through contributions of all authors. All authors have given approval to the final version of the manuscript.

Acknowledgments: This work was supported by the National Natural Science Foundation of China (Grant Nos. 61504048, 51675517, and 61505241), Natural Science Fund of Jiangsu Province (Grant Nos. BK20140380 and BK20160057), Youth Innovation Promotion Association Chinese Academy of Sciences (Grant No. 2014280), Ministry of Science and Technology focused on Special International Cooperation Projects (Grant No. 2016YFE0107700), Jiangsu Planned Projects for Postdoctoral Research Funds (Grant No. 1701010A), Major Technology Innovation Projects of Jiangsu Province (Grant No. BO2015007), China Postdoctoral Science Foundation (Grant No. 2016M601890), Shenzhen Key Laboratory Project (Grant No. ZDSYS201603311644527), Shenzhen Fundamental Research Fund (Grant Nos. JCYJ20150611092848134 and JCYJ20150929170644623), and the Shenzhen Science and Technology Innovation Fund (Grant No. KQCX20140522143114399).

Conflicts of Interest: The authors declare no conflict of interest.

References

1. Wang, Z.L. Zinc oxide nanostructures: Growth, properties and applications. *J. Phys. Condens. Matter* **2004**, *16*, R829–R858. [CrossRef]
2. Vayssieres, L. Growth of arrayed nanorods and nanowires of ZnO from aqueous solutions. *Adv. Mater.* **2003**, *15*, 464–466. [CrossRef]
3. Greene, L.E.; Law, M.; Tan, D.H.; Montano, M.; Goldberger, J.; Somorjai, G.; Yang, P. General route to vertical ZnO nanowire arrays using textured ZnO seeds. *Nano Lett.* **2005**, *5*, 1231–1236. [CrossRef] [PubMed]
4. Soci, C.; Zhang, A.; Xiang, B.; Dayeh, S.A.; Aplin, D.P.R.; Park, J.; Bao, X.Y.; Lo, Y.H.; Wang, D. ZnO Nanowire UV Photodetectors with High Internal Gain. *Nano Lett.* **2007**, *7*, 1003–1009. [CrossRef] [PubMed]
5. Huang, M.H.; Wu, Y.; Feick, H.; Tran, N.; Weber, E.; Yang, P. Catalytic Growth of Zinc Oxide Nanowires by Vapor Transport. *Adv. Mater.* **2001**, *13*, 113–116. [CrossRef]
6. Fang, F.; Zhao, D.X.; Zhang, J.Y.; Shen, D.Z.; Lu, Y.M.; Fan, X.W.; Li, B.H.; Wang, X.H. Growth of well-aligned ZnO nanowire arrays on Si substrate. *Nanotechnology* **2007**, *18*, 235604. [CrossRef]
7. Jung, S.H.; Oh, E.; Lee, K.H.; Park, W.; Jeong, S.H. A Sonochemical Method for Fabricating Aligned ZnO Nanorods. *Adv. Mater.* **2007**, *19*, 749–753. [CrossRef]
8. Kim, K.S.; Jeong, H.; Jeong, M.S.; Jung, G.Y. Polymer-Templated Hydrothermal Growth of Vertically Aligned Single-Crystal ZnO Nanorods and Morphological Transformations Using Structural Polarity. *Adv. Funct. Mater.* **2010**, *20*, 3055–3063. [CrossRef]
9. Liu, B.; Zeng, H.C. Hydrothermal Synthesis of ZnO Nanorods in the Diameter Regime of 50 nm. *J. Am. Chem. Soc.* **2003**, *125*, 4430–4431. [CrossRef] [PubMed]
10. Wang, W.Z.; Zeng, B.Q.; Yang, J.; Poudel, B.; Huang, J.Y.; Naughton, M.J.; Ren, Z.F. Aligned Ultralong ZnO Nanobelts and Their Enhanced Field Emission. *Adv. Mater.* **2006**, *18*, 3275–3278. [CrossRef]
11. Lu, C.; Qi, L.; Yang, J.; Tang, L.; Zhang, D.; Ma, J. Hydrothermal growth of large-scale micropatterned arrays of ultralong ZnO nanowires and nanobelts on zinc substrate. *Chem. Commun.* **2006**, *0*, 3551–3553. [CrossRef] [PubMed]
12. Yang, J.; Liu, G.; Lu, J.; Qiu, Y.; Yang, S. Electrochemical route to the synthesis of ultrathin ZnO nanorod/nanobelt arrays on zinc substrate. *Appl. Phys. Lett.* **2007**, *90*, 103109. [CrossRef]
13. Wang, X.; Ding, Y.; Summers, C.J.; Wang, Z.L. Large-Scale Synthesis of Six-Nanometer-Wide ZnO Nanobelts. *J. Phys. Chem. B* **2004**, *108*, 8773–8777. [CrossRef]
14. Sun, Y.; Fuge, G.M.; Fox, N.A.; Riley, D.J.; Ashfold, M.N.R. Synthesis of Aligned Arrays of Ultrathin ZnO Nanotubes on a Si Wafer Coated with a Thin ZnO Film. *Adv. Mater.* **2005**, *17*, 2477–2481. [CrossRef]

15. Xing, Y.J.; Xi, Z.H.; Xue, Z.Q.; Zhang, X.D.; Song, J.H.; Wang, R.M.; Xu, J.; Song, Y.; Zhang, S.L.; Yu, D.P. Optical properties of the ZnO nanotubes synthesized via vapor phase growth. *Appl. Phys. Lett.* **2003**, *83*, 1689–1691. [CrossRef]
16. Wei, A.; Sun, X.W.; Xu, C.X.; Dong, Z.L.; Yu, M.B.; Huang, W. Stable field emission from hydrothermally grown ZnO nanotubes. *Appl. Phys. Lett.* **2006**, *88*, 213102. [CrossRef]
17. Reimer, T.; Paulowicz, I.; Röder, R.; Kaps, S.; Lupan, O.; Chemnitz, S.; Benecke, W.; Ronning, C.; Adelung, R.; Mishra, Y.K. Single Step Integration of ZnO Nano- and Microneedles in Si Trenches by Novel Flame Transport Approach: Whispering Gallery Modes and Photocatalytic Properties. *ACS Appl. Mater. Interfaces* **2014**, *6*, 7806–7815. [CrossRef] [PubMed]
18. Kaps, S.; Bhowmick, S.; Gröttrup, J.; Hrkac, V.; Stauffer, D.; Guo, H.; Warren, O.L.; Adam, J.; Kienle, L.; Minor, A.M.; et al. Piezoresistive Response of Quasi-One-Dimensional ZnO Nanowires Using an in Situ Electromechanical Device. *ACS Omega* **2017**, *2*, 2985–2993. [CrossRef]
19. Zhang, C.; Zhu, F.; Xu, H.; Liu, W.; Yang, L.; Wang, Z.; Ma, J.; Kang, Z.; Liu, Y. Significant improvement of near-UV electroluminescence from ZnO quantum dot LEDs via coupling with carbon nanodot surface plasmons. *Nanoscale* **2017**, *9*, 14592–14601. [CrossRef] [PubMed]
20. Mishra, Y.K.; Modi, G.; Cretu, V.; Postica, V.; Lupan, O.; Reimer, T.; Paulowicz, I.; Hrkac, V.; Benecke, W.; Kienle, L.; et al. Direct Growth of Freestanding ZnO Tetrapod Networks for Multifunctional Applications in Photocatalysis, UV Photodetection, and Gas Sensing. *ACS Appl. Mater. Interfaces* **2015**, *7*, 14303–14316. [CrossRef] [PubMed]
21. Özgür, Ü.; Alivov, Y.I.; Liu, C.; Teke, A.; Reshchikov, M.A.; Doğan, S.; Avrutin, V.; Cho, S.-J.; Morkoç, H. A comprehensive review of ZnO materials and devices. *J. Appl. Phys.* **2005**, *98*, 41301. [CrossRef]
22. Teng, M.; Min, G.; Mei, Z.; Yanjun, Z.; Xidong, W. Density-controlled hydrothermal growth of well-aligned ZnO nanorod arrays. *Nanotechnology* **2007**, *18*, 35605. [CrossRef]
23. Li, L.; Pan, S.; Dou, X.; Zhu, Y.; Huang, X.; Yang, Y.; Li, G.; Zhang, L. Direct Electrodeposition of ZnO Nanotube Arrays in Anodic Alumina Membranes. *J. Phys. Chem. C* **2007**, *111*, 7288–7291. [CrossRef]
24. Wang, Z.; Liu, X.; Gong, J.; Huang, H.; Gu, S.; Yang, S. Epitaxial Growth of ZnO Nanowires on ZnS Nanobelts by Metal Organic Chemical Vapor Deposition. *Cryst. Growth Des.* **2008**, *8*, 3911–3913. [CrossRef]
25. Woong, L.; Min-Chang, J.; Jae-Min, M. Fabrication and application potential of ZnO nanowires grown on GaAs(002) substrates by metal–organic chemical vapour deposition. *Nanotechnology* **2004**, *15*, 254. [CrossRef]
26. Heo, Y.W.; Varadarajan, V.; Kaufman, M.; Kim, K.; Norton, D.P.; Ren, F.; Fleming, P.H. Site-specific growth of Zno nanorods using catalysis-driven molecular-beam epitaxy. *Appl. Phys. Lett.* **2002**, *81*, 3046–3048. [CrossRef]
27. Cao, B.Q.; Lorenz, M.; Rahm, A.; Wenckstern, H.v.; Czekalla, C.; Lenzner, J.; Benndorf, G.; Grundmann, M. Phosphorus acceptor doped ZnO nanowires prepared by pulsed-laser deposition. *Nanotechnology* **2007**, *18*, 455707. [CrossRef]
28. Acuña, K.; Yáñez, J.; Ranganathan, S.; Ramírez, E.; Pablo Cuevas, J.; Mansilla, H.D.; Santander, P. Photocatalytic degradation of roxarsone by using synthesized ZnO nanoplates. *Sol. Energy* **2017**, *157*, 335–341. [CrossRef]
29. Xu, F.; Shen, Y.; Sun, L.; Zeng, H.; Lu, Y. Enhanced photocatalytic activity of hierarchical ZnO nanoplate-nanowire architecture as environmentally safe and facilely recyclable photocatalyst. *Nanoscale* **2011**, *3*, 5020–5025. [CrossRef]
30. Song, J.; Kulinich, S.A.; Yan, J.; Li, Z.; He, J.; Kan, C.; Zeng, H. Epitaxial ZnO Nanowire-on-Nanoplate Structures as Efficient and Transferable Field Emitters. *Adv. Mater.* **2013**, *25*, 5750–5755. [CrossRef] [PubMed]
31. Weng, B.; Yang, M.-Q.; Zhang, N.; Xu, Y.-J. Toward the enhanced photoactivity and photostability of ZnO nanospheres via intimate surface coating with reduced graphene oxide. *J. Mater. Chem. A* **2014**, *2*, 9380–9389. [CrossRef]
32. Qiu, Y.; Yang, D.C.; Yin, B.; Lei, J.X.; Zhang, H.Q.; Zhang, Z.; Chen, H.; Li, Y.P.; Bian, J.M.; Liu, Y.H.; et al. Branched ZnO nanotrees on flexible fiber-paper substrates for self-powered energy-harvesting systems. *RSC Adv.* **2015**, *5*, 5941–5945. [CrossRef]
33. Cheng, L.; Chang, Q.; Chang, Y.; Zhang, N.; Tao, C.; Wang, Z.; Fan, X. Hierarchical forest-like photoelectrodes with ZnO nanoleaves on a metal dendrite array. *J. Mater. Chem. A* **2016**, *4*, 9816–9821. [CrossRef]

34. Jebril, S.; Kuhlmann, H.; Müller, S.; Ronning, C.; Kienle, L.; Duppel, V.; Mishra, Y.K.; Adelung, R. Epitactically Interpenetrated High Quality ZnO Nanostructured Junctions on Microchips Grown by the Vapor−Liquid−Solid Method. *Cryst. Growth Des.* **2010**, *10*, 2842–2846. [CrossRef]
35. Mishra, Y.K.; Adelung, R. ZnO tetrapod materials for functional applications. *Mater. Today* **2017**. [CrossRef]
36. Hrkac, S.B.; Koops, C.T.; Abes, M.; Krywka, C.; Müller, M.; Burghammer, M.; Sztucki, M.; Dane, T.; Kaps, S.; Mishra, Y.K.; et al. Tunable Strain in Magnetoelectric ZnO Microrod Composite Interfaces. *ACS Appl. Mater. Interfaces* **2017**, *9*, 25571–25577. [CrossRef] [PubMed]
37. Ding, M.; Zhao, D.; Yao, B.; E, S.; Guo, Z.; Zhang, L.; Shen, D. The ultraviolet laser from individual ZnO microwire with quadrate cross section. *Opt. Express* **2012**, *20*, 13657–13662. [CrossRef] [PubMed]
38. Li, Z.; Zhong, W.; Li, X.; Zeng, H.; Wang, G.; Wang, W.; Yang, Z.; Zhang, Y. Strong room-temperature ferromagnetism of pure ZnO nanostructure arrays via colloidal template. *J. Mater. Chem. C* **2013**, *1*, 6807–6812. [CrossRef]
39. Mishra, Y.K.; Mohapatra, S.; Singhal, R.; Avasthi, D.K.; Agarwal, D.C.; Ogale, S.B. Au–ZnO: A tunable localized surface plasmonic nanocomposite. *Appl. Phys. Lett.* **2008**, *92*, 43107. [CrossRef]
40. Mishra, Y.K.; Sören, K.; Arnim, S.; Ingo, P.; Xin, J.; Dawit, G.; Stefan, F.; Maria, C.; Sebastian, W.; Alexander, K.; et al. Fabrication of Macroscopically Flexible and Highly Porous 3D Semiconductor Networks from Interpenetrating Nanostructures by a Simple Flame Transport Approach. *Part. Part. Syst. Charact.* **2013**, *30*, 775–783. [CrossRef]
41. Yi, G.-C.; Wang, C.; Park, W.I. ZnO nanorods: Synthesis, characterization and applications. *Semiconduct. Sci. Technol.* **2005**, *20*, S22. [CrossRef]
42. Xu, S.; Wang, Z.L. One-dimensional ZnO nanostructures: Solution growth and functional properties. *Nano Res.* **2011**, *4*, 1013–1098. [CrossRef]
43. Reeber, R.R. Lattice parameters of ZnO from 4.2° to 296°K. *J. Appl. Phys.* **1970**, *41*, 5063–5066. [CrossRef]
44. Guo, Z.; Zhao, D.; Liu, Y.; Shen, D.; Zhang, J.; Li, B. Visible and ultraviolet light alternative photodetector based on ZnO nanowire/n-Si heterojunction. *Appl. Phys. Lett.* **2008**, *93*, 163501. [CrossRef]
45. Xu, C.X.; Sun, X.W. Field emission from zinc oxide nanopins. *Appl. Phys. Lett.* **2003**, *83*, 3806–3808. [CrossRef]
46. Gao, P.X.; Ding, Y.; Wang, Z.L. Crystallographic Orientation-Aligned ZnO Nanorods Grown by a Tin Catalyst. *Nano Lett.* **2003**, *3*, 1315–1320. [CrossRef]
47. Lee, C.J.; Lee, T.J.; Lyu, S.C.; Zhang, Y.; Ruh, H.; Lee, H.J. Field emission from well-aligned zinc oxide nanowires grown at low temperature. *Appl. Phys. Lett.* **2002**, *81*, 3648–3650. [CrossRef]
48. Xu, X.Y.; Zhang, H.Z.; Zhao, Q.; Chen, Y.F.; Xu, J.; Yu, D.P. Patterned Growth of ZnO Nanorod Arrays on a Large-Area Stainless Steel Grid. *J. Phys. Chem. B* **2005**, *109*, 1699–1702. [CrossRef] [PubMed]
49. Zhu, Z.; Chen, T.-L.; Gu, Y.; Warren, J.; Osgood, R.M. Zinc Oxide Nanowires Grown by Vapor-Phase Transport Using Selected Metal Catalysts: A Comparative Study. *Chem. Mater.* **2005**, *17*, 4227–4234. [CrossRef]
50. Yang, Y.H.; Wang, C.X.; Wang, B.; Li, Z.Y.; Chen, J.; Chen, D.H.; Xu, N.S.; Yang, G.W.; Xu, J.B. Radial ZnO nanowire nucleation on amorphous carbons. *Appl. Phys. Lett.* **2005**, *87*, 183109. [CrossRef]
51. Vayssieres, L.; Keis, K.; Lindquist, S.-E.; Hagfeldt, A. Purpose-Built Anisotropic Metal Oxide Material: 3D Highly Oriented Microrod Array of ZnO. *J. Phys. Chem. B* **2001**, *105*, 3350–3352. [CrossRef]
52. Greene, L.E.; Law, M.; Goldberger, J.; Kim, F.; Johnson, J.C.; Zhang, Y.; Saykally, R.J.; Yang, P. Low-Temperature Wafer-Scale Production of ZnO Nanowire Arrays. *Angew. Chem. Int. Ed.* **2003**, *42*, 3031–3034. [CrossRef] [PubMed]
53. Chen, S.-W.; Wu, J.-M. Nucleation mechanisms and their influences on characteristics of ZnO nanorod arrays prepared by a hydrothermal method. *Acta Mater.* **2011**, *59*, 841–847. [CrossRef]
54. Manekkathodi, A.; Lu, M.-Y.; Wang, C.W.; Chen, L.-J. Direct Growth of Aligned Zinc Oxide Nanorods on Paper Substrates for Low-Cost Flexible Electronics. *Adv. Mater.* **2010**, *22*, 4059–4063. [CrossRef] [PubMed]
55. Na, J.-S.; Gong, B.; Scarel, G.; Parsons, G.N. Surface Polarity Shielding and Hierarchical ZnO Nano-Architectures Produced Using Sequential Hydrothermal Crystal Synthesis and Thin Film Atomic Layer Deposition. *ACS Nano* **2009**, *3*, 3191–3199. [CrossRef] [PubMed]
56. Greene, L.E.; Yuhas, B.D.; Law, M.; Zitoun, D.; Yang, P. Solution-Grown Zinc Oxide Nanowires. *Inorg. Chem.* **2006**, *45*, 7535–7543. [CrossRef] [PubMed]
57. Sun, X.M.; Chen, X.; Deng, Z.X.; Li, Y.D. A CTAB-assisted hydrothermal orientation growth of ZnO nanorods. *Mater. Chem. Phys.* **2003**, *78*, 99–104. [CrossRef]

58. Zeng, H.; Cui, J.; Cao, B.; Gibson, U.; Bando, Y.; Golberg, D. Electrochemical Deposition of ZnO Nanowire Arrays: Organization, Doping, and Properties. *Sci. Adv. Mater.* **2010**, *2*, 336–358. [CrossRef]
59. Weintraub, B.; Zhou, Z.; Li, Y.; Deng, Y. Solution synthesis of one-dimensional ZnO nanomaterials and their applications. *Nanoscale* **2010**, *2*, 1573–1587. [CrossRef] [PubMed]
60. Tang, Y.; Luo, L.; Chen, Z.; Jiang, Y.; Li, B.; Jia, Z.; Xu, L. Electrodeposition of ZnO nanotube arrays on TCO glass substrates. *Electrochem. Commun.* **2007**, *9*, 289–292. [CrossRef]
61. Khajavi, M.R.; Blackwood, D.J.; Cabanero, G.; Tena-Zaera, R. New insight into growth mechanism of ZnO nanowires electrodeposited from nitrate-based solutions. *Electrochim. Acta* **2012**, *69*, 181–189. [CrossRef]
62. Xu, L.; Guo, Y.; Liao, Q.; Zhang, J.; Xu, D. Morphological Control of ZnO Nanostructures by Electrodeposition. *J. Phys. Chem. B* **2005**, *109*, 13519–13522. [CrossRef] [PubMed]
63. Wong, M.H.; Berenov, A.; Qi, X.; Kappers, M.J.; Barber, Z.H.; Illy, B.; Lockman, Z.; Ryan, M.P.; MacManus-Driscoll, J.L. Electrochemical growth of ZnO nano-rods on polycrystalline Zn foil. *Nanotechnology* **2003**, *14*, 968. [CrossRef]
64. Cao, B.; Li, Y.; Duan, G.; Cai, W. Growth of ZnO Nanoneedle Arrays with Strong Ultraviolet Emissions by an Electrochemical Deposition Method. *Cryst. Growth Des.* **2006**, *6*, 1091–1095. [CrossRef]
65. Marí, B.; Mollar, M.; Mechkour, A.; Hartiti, B.; Perales, M.; Cembrero, J. Optical properties of nanocolumnar ZnO crystals. *Microelectron. J.* **2004**, *35*, 79–82. [CrossRef]
66. Tena-Zaera, R.; Elias, J.; Wang, G.; Lévy-Clément, C. Role of Chloride Ions on Electrochemical Deposition of ZnO Nanowire Arrays from O2 Reduction. *J. Phys. Chem. C* **2007**, *111*, 16706–16711. [CrossRef]
67. Elias, J.; Tena-Zaera, R.; Lévy-Clément, C. Effect of the Chemical Nature of the Anions on the Electrodeposition of ZnO Nanowire Arrays. *J. Phys. Chem. C* **2008**, *112*, 5736–5741. [CrossRef]
68. Yi, J.B.; Pan, H.; Lin, J.Y.; Ding, J.; Feng, Y.P.; Thongmee, S.; Liu, T.; Gong, H.; Wang, L. Ferromagnetism in ZnO Nanowires Derived from Electro-deposition on AAO Template and Subsequent Oxidation. *Adv. Mater.* **2008**, *20*, 1170–1174. [CrossRef]
69. Zheng, M.J.; Zhang, L.D.; Li, G.H.; Shen, W.Z. Fabrication and optical properties of large-scale uniform zinc oxide nanowire arrays by one-step electrochemical deposition technique. *Chem. Phys. Lett* **2002**, *363*, 123–128. [CrossRef]
70. Li, Y.; Meng, G.W.; Zhang, L.D.; Phillipp, F. Ordered semiconductor ZnO nanowire arrays and their photoluminescence properties. *Appl. Phys. Lett.* **2000**, *76*, 2011–2013. [CrossRef]
71. Leprince-Wang, Y.; Bouchaib, S.; Brouri, T.; Capo-Chichi, M.; Laurent, K.; Leopoldes, J.; Tusseau-Nenez, S.; Lei, L.; Chen, Y. Fabrication of ZnO micro- and nano-structures by electrodeposition using nanoporous and lithography defined templates. *Mater. Sci. Eng. B* **2010**, *170*, 107–112. [CrossRef]
72. Zhou, H.; Wong, S.S. A Facile and Mild Synthesis of 1-D ZnO, CuO, and α-Fe2O3 Nanostructures and Nanostructured Arrays. *ACS Nano* **2008**, *2*, 944–958. [CrossRef] [PubMed]
73. Song, J.; Ning, X.; Zeng, H. ZnO nanowire lines and bundles: Template-deformation-guided alignment for patterned field-electron emitters. *Curr. Appl. Phys.* **2015**, *15*, 1296–1302. [CrossRef]
74. Sun, X.W.; Ling, B.; Zhao, J.L.; Tan, S.T.; Yang, Y.; Shen, Y.Q.; Dong, Z.L.; Li, X.C. Ultraviolet emission from a ZnO rod homojunction light-emitting diode. *Appl. Phys. Lett.* **2009**, *95*, 133124. [CrossRef]
75. Chang, C.-Y.; Tsao, F.-C.; Pan, C.-J.; Chi, G.-C.; Wang, H.-T.; Chen, J.-J.; Ren, F.; Norton, D.P.; Pearton, S.J.; Chen, K.-H.; et al. Electroluminescence from ZnO nanowire/polymer composite p-n junction. *Appl. Phys. Lett.* **2006**, *88*, 173503. [CrossRef]
76. Zhang, X.-M.; Lu, M.-Y.; Zhang, Y.; Chen, L.-J.; Wang, Z.L. Fabrication of a High-Brightness Blue-Light-Emitting Diode Using a ZnO-Nanowire Array Grown on p-GaN Thin Film. *Adv. Mater.* **2009**, *21*, 2767–2770. [CrossRef]
77. Tang, X.; Li, G.; Zhou, S. Ultraviolet Electroluminescence of Light-Emitting Diodes Based on Single n-ZnO/p-AlGaN Heterojunction Nanowires. *Nano Lett.* **2013**, *13*, 5046–5050. [CrossRef] [PubMed]
78. Lupan, O.; Pauporté, T.; Viana, B. Low-Voltage UV-Electroluminescence from ZnO-Nanowire Array/p-GaN Light-Emitting Diodes. *Adv. Mater.* **2010**, *22*, 3298–3302. [CrossRef] [PubMed]
79. Lupan, O.; Pauporté, T.; Viana, B.; Tiginyanu, I.M.; Ursaki, V.V.; Cortès, R. Epitaxial Electrodeposition of ZnO Nanowire Arrays on p-GaN for Efficient UV-Light-Emitting Diode Fabrication. *ACS Appl. Mater. Interfaces* **2010**, *2*, 2083–2090. [CrossRef]
80. An, S.J.; Yi, G.-C. Near ultraviolet light emitting diode composed of n-GaN/ZnO coaxial nanorod heterostructures on a p-GaN layer. *Appl. Phys. Lett.* **2007**, *91*, 123109. [CrossRef]

81. Lee, S.W.; Cho, H.D.; Panin, G.; Kang, T.W. Vertical ZnO nanorod/Si contact light-emitting diode. *Appl. Phys. Lett.* **2011**, *98*, 93110. [CrossRef]
82. Zimmler, M.A.; Voss, T.; Ronning, C.; Capasso, F. Exciton-related electroluminescence from ZnO nanowire light-emitting diodes. *Appl. Phys. Lett.* **2009**, *94*, 241120. [CrossRef]
83. Liu, X.-Y.; Shan, C.-X.; Jiao, C.; Wang, S.-P.; Zhao, H.-F.; Shen, D.-Z. Pure ultraviolet emission from ZnO nanowire-based *p*-*n* heterostructures. *Opt. Lett.* **2014**, *39*, 422–425. [CrossRef] [PubMed]
84. Wang, J.-Y.; Lee, C.-Y.; Chen, Y.-T.; Chen, C.-T.; Chen, Y.-L.; Lin, C.-F.; Chen, Y.-F. Double side electroluminescence from *p*-NiO/*n*-ZnO nanowire heterojunctions. *Appl. Phys. Lett.* **2009**, *95*, 131117. [CrossRef]
85. Sun, X.W.; Huang, J.Z.; Wang, J.X.; Xu, Z. A ZnO Nanorod Inorganic/Organic Heterostructure Light-Emitting Diode Emitting at 342 nm. *Nano Lett.* **2008**, *8*, 1219–1223. [CrossRef] [PubMed]
86. Zhao, S.-L.; Kan, P.-Z.; Xu, Z.; Kong, C.; Wang, D.-W.; Yan, Y.; Wang, Y.-S. Electroluminescence of ZnO nanorods/MEH-PPV heterostructure devices. *Org. Electron.* **2010**, *11*, 789–793. [CrossRef]
87. Ren, X.; Zhang, X.; Liu, N.; Wen, L.; Ding, L.; Ma, Z.; Su, J.; Li, L.; Han, J.; Gao, Y. White Light-Emitting Diode From Sb-Doped *p*-ZnO Nanowire Arrays/*n*-GaN Film. *Adv. Funct. Mater.* **2015**, *25*, 2182–2188. [CrossRef]
88. Sun, H.; Zhang, Q.-F.; Wu, J.-L. Electroluminescence from ZnO nanorods with an *n*-ZnO/*p*-Si heterojunction structure. *Nanotechnology* **2006**, *17*, 2271. [CrossRef]
89. Huang, J.; Chu, S.; Kong, J.; Zhang, L.; Schwarz, C.M.; Wang, G.; Chernyak, L.; Chen, Z.; Liu, J. ZnO *p*–*n* Homojunction Random Laser Diode Based on Nitrogen-Doped p-type Nanowires. *Adv. Opt. Mater.* **2013**, *1*, 179–185. [CrossRef]
90. Zhang, J.-Y.; Zhang, Q.-F.; Deng, T.-S.; Wu, J.-L. Electrically driven ultraviolet lasing behavior from phosphorus-doped *p*-ZnO nanonail array/*n*-Si heterojunction. *Appl. Phys. Lett.* **2009**, *95*, 211107. [CrossRef]
91. Zhang, J.-Y.; Li, P.-J.; Sun, H.; Shen, X.; Deng, T.-S.; Zhu, K.-T.; Zhang, Q.-F.; Wu, J.-L. Ultraviolet electroluminescence from controlled arsenic-doped ZnO nanowire homojunctions. *Appl. Phys. Lett.* **2008**, *93*, 21116. [CrossRef]
92. Yang, Y.; Sun, X.W.; Tay, B.K.; You, G.F.; Tan, S.T.; Teo, K.L. A p–n homojunction ZnO nanorod light-emitting diode formed by As ion implantation. *Appl. Phys. Lett.* **2008**, *93*, 253107. [CrossRef]
93. Chu, S.; Wang, G.; Zhou, W.; Lin, Y.; Chernyak, L.; Zhao, J.; Kong, J.; Li, L.; Ren, J.; Liu, J. Electrically pumped waveguide lasing from ZnO nanowires. *Nat. Nanotechnol.* **2011**, *6*, 506–510. [CrossRef] [PubMed]
94. Gao, F.; Zhang, D.; Wang, J.; Sun, H.; Yin, Y.; Sheng, Y.; Yan, S.; Yan, B.; Sui, C.; Zheng, Y.; et al. Ultraviolet electroluminescence from Au-ZnO nanowire Schottky type light-emitting diodes. *Appl. Phys. Lett.* **2016**, *108*, 261103. [CrossRef]
95. Liu, W.Z.; Xu, H.Y.; Ma, J.G.; Liu, C.Y.; Liu, Y.X.; Liu, Y.C. Effect of oxygen-related surface adsorption on the efficiency and stability of ZnO nanorod array ultraviolet light-emitting diodes. *Appl. Phys. Lett.* **2012**, *100*, 203101. [CrossRef]
96. Liu, W.; Xu, H.; Yan, S.; Zhang, C.; Wang, L.; Wang, C.; Yang, L.; Wang, X.; Zhang, L.; Wang, J.; et al. Effect of SiO2 Spacer-Layer Thickness on Localized Surface Plasmon-Enhanced ZnO Nanorod Array LEDs. *ACS Appl. Mater. Interfaces* **2016**, *8*, 1653–1660. [CrossRef] [PubMed]
97. Liu, C.Y.; Xu, H.Y.; Ma, J.G.; Li, X.H.; Zhang, X.T.; Liu, Y.C.; Mu, R. Electrically pumped near-ultraviolet lasing from ZnO/MgO core/shell nanowires. *Appl. Phys. Lett.* **2011**, *99*, 63115. [CrossRef]
98. Liu, W.Z.; Xu, H.Y.; Wang, C.L.; Zhang, L.X.; Zhang, C.; Sun, S.Y.; Ma, J.G.; Zhang, X.T.; Wang, J.N.; Liu, Y.C. Enhanced ultraviolet emission and improved spatial distribution uniformity of ZnO nanorod array light-emitting diodes via Ag nanoparticles decoration. *Nanoscale* **2013**, *5*, 8634–8639. [CrossRef] [PubMed]
99. Lu, J.; Shi, Z.; Wang, Y.; Lin, Y.; Zhu, Q.; Tian, Z.; Dai, J.; Wang, S.; Xu, C. Plasmon-enhanced Electrically Light-emitting from ZnO Nanorod Arrays/*p*-GaN Heterostructure Devices. *Sci. Rep.* **2016**, *6*, 25645. [CrossRef] [PubMed]
100. Zhang, C.; Marvinney, C.E.; Xu, H.Y.; Liu, W.Z.; Wang, C.L.; Zhang, L.X.; Wang, J.N.; Ma, J.G.; Liu, Y.C. Enhanced waveguide-type ultraviolet electroluminescence from ZnO/MgZnO core/shell nanorod array light-emitting diodes via coupling with Ag nanoparticles localized surface plasmons. *Nanoscale* **2015**, *7*, 1073–1080. [CrossRef] [PubMed]
101. Wang, D.-W.; Zhao, S.-L.; Xu, Z.; Kong, C.; Gong, W. The improvement of near-ultraviolet electroluminescence of ZnO nanorods/MEH-PPV heterostructure by using a ZnS buffer layer. *Org. Electron.* **2011**, *12*, 92–97. [CrossRef]

102. Fang, X.; Wei, Z.; Yang, Y.; Chen, R.; Li, Y.; Tang, J.; Fang, D.; Jia, H.; Wang, D.; Fan, J.; et al. Ultraviolet Electroluminescence from ZnS@ZnO Core–Shell Nanowires/*p*-GaN Introduced by Exciton Localization. *ACS Appl. Mater. Interfaces* **2016**, *8*, 1661–1666. [CrossRef] [PubMed]
103. Wang, C.; Bao, R.; Zhao, K.; Zhang, T.; Dong, L.; Pan, C. Enhanced emission intensity of vertical aligned flexible ZnO nanowire/p-polymer hybridized LED array by piezo-phototronic effect. *Nano Energy* **2015**, *14*, 364–371. [CrossRef]
104. Yang, Q.; Liu, Y.; Pan, C.; Chen, J.; Wen, X.; Wang, Z.L. Largely Enhanced Efficiency in ZnO Nanowire/p-Polymer Hybridized Inorganic/Organic Ultraviolet Light-Emitting Diode by Piezo-Phototronic Effect. *Nano Lett.* **2013**, *13*, 607–613. [CrossRef] [PubMed]
105. Sharma, P.; Sreenivas, K.; Rao, K.V. Analysis of ultraviolet photoconductivity in ZnO films prepared by unbalanced magnetron sputtering. *J. Appl. Phys.* **2003**, *93*, 3963–3970. [CrossRef]
106. Peng, L.; Hu, L.; Fang, X. Low-Dimensional Nanostructure Ultraviolet Photodetectors. *Adv. Mater.* **2013**, *25*, 5321–5328. [CrossRef] [PubMed]
107. Cheng, G.; Wu, X.; Liu, B.; Li, B.; Zhang, X.; Du, Z. ZnO nanowire Schottky barrier ultraviolet photodetector with high sensitivity and fast recovery speed. *Appl. Phys. Lett.* **2011**, *99*, 203105. [CrossRef]
108. Flemban, T.H.; Haque, M.A.; Ajia, I.; Alwadai, N.; Mitra, S.; Wu, T.; Roqan, I.S. A Photodetector Based on *p*-Si/*n*-ZnO Nanotube Heterojunctions with High Ultraviolet Responsivity. *ACS Appl. Mater. Interfaces* **2017**, *9*, 37120–37127. [CrossRef] [PubMed]
109. Cho, H.D.; Zakirov, A.S.; Yuldashev, S.U.; Ahn, C.W.; Yeo, Y.K.; Kang, T.W. Photovoltaic device on a single ZnO nanowire *p–n* homojunction. *Nanotechnology* **2012**, *23*, 115401. [CrossRef] [PubMed]
110. Wang, L.; Zhao, D.; Su, Z.; Fang, F.; Li, B.; Zhang, Z.; Shen, D.; Wang, X. High spectrum selectivity organic/inorganic hybrid visible-blind ultraviolet photodetector based on ZnO nanorods. *Org. Electron.* **2010**, *11*, 1318–1322. [CrossRef]
111. Kind, H.; Yan, H.; Messer, B.; Law, M.; Yang, P. Nanowire Ultraviolet Photodetectors and Optical Switches. *Adv. Mater.* **2002**, *14*, 158–160. [CrossRef]
112. Hu, Y.; Zhou, J.; Yeh, P.-H.; Li, Z.; Wei, T.-Y.; Wang, Z.L. Supersensitive, Fast-Response Nanowire Sensors by Using Schottky Contacts. *Adv. Mater.* **2010**, *22*, 3327–3332. [CrossRef] [PubMed]
113. Zhou, J.; Gu, Y.; Hu, Y.; Mai, W.; Yeh, P.-H.; Bao, G.; Sood, A.K.; Polla, D.L.; Wang, Z.L. Gigantic enhancement in response and reset time of ZnO UV nanosensor by utilizing Schottky contact and surface functionalization. *Appl. Phys. Lett.* **2009**, *94*, 191103. [CrossRef] [PubMed]
114. Fu, X.-W.; Liao, Z.-M.; Zhou, Y.-B.; Wu, H.-C.; Bie, Y.-Q.; Xu, J.; Yu, D.-P. Graphene/ZnO nanowire/graphene vertical structure based fast-response ultraviolet photodetector. *Appl. Phys. Lett.* **2012**, *100*, 223114. [CrossRef]
115. Nie, B.; Hu, J.-G.; Luo, L.-B.; Xie, C.; Zeng, L.-H.; Lv, P.; Li, F.-Z.; Jie, J.-S.; Feng, M.; Wu, C.-Y.; et al. Monolayer Graphene Film on ZnO Nanorod Array for High-Performance Schottky Junction Ultraviolet Photodetectors. *Small* **2013**, *9*, 2872–2879. [CrossRef] [PubMed]
116. Panigrahi, S.; Basak, D. Solution-processed novel core-shell *n-p* heterojunction and its ultrafast UV photodetection properties. *RSC Adv.* **2012**, *2*, 11963–11968. [CrossRef]
117. Leung, Y.H.; He, Z.B.; Luo, L.B.; Tsang, C.H.A.; Wong, N.B.; Zhang, W.J.; Lee, S.T. ZnO nanowires array *p-n* homojunction and its application as a visible-blind ultraviolet photodetector. *Appl. Phys. Lett.* **2010**, *96*, 53102. [CrossRef]
118. Hatch, S.M.; Briscoe, J.; Dunn, S. A Self-Powered ZnO-Nanorod/CuSCN UV Photodetector Exhibiting Rapid Response. *Adv. Mater.* **2013**, *25*, 867–871. [CrossRef] [PubMed]
119. Wang, Y.; Chen, Y.; Zhao, W.; Ding, L.; Wen, L.; Li, H.; Jiang, F.; Su, J.; Li, L.; Liu, N.; et al. A Self-Powered Fast-Response Ultraviolet Detector of *p-n* Homojunction Assembled from Two ZnO-Based Nanowires. *Nano-Micro Lett.* **2016**, *9*, 11. [CrossRef]
120. Bie, Y.Q.; Liao, Z.M.; Zhang, H.Z.; Li, G.R.; Ye, Y.; Zhou, Y.B.; Xu, J.; Qin, Z.X.; Dai, L.; Yu, D.P. Self-Powered, Ultrafast, Visible-Blind UV Detection and Optical Logical Operation based on ZnO/GaN Nanoscale *p-n* Junctions. *Adv. Mater.* **2011**, *23*, 649–653. [CrossRef] [PubMed]
121. Zhou, H.; Gui, P.; Yu, Q.; Mei, J.; Wang, H.; Fang, G. Self-powered, visible-blind ultraviolet photodetector based on n-ZnO nanorods/i-MgO/p-GaN structure light-emitting diodes. *J. Mater. Chem. C* **2015**, *3*, 990–994. [CrossRef]

122. Huang, H.; Fang, G.; Mo, X.; Yuan, L.; Zhou, H.; Wang, M.; Xiao, H.; Zhao, X. Zero-biased near-ultraviolet and visible photodetector based on ZnO nanorods/n-Si heterojunction. *Appl. Phys. Lett.* **2009**, *94*, 63512. [CrossRef]

123. Garnier, J.; Parize, R.; Appert, E.; Chaix-Pluchery, O.; Kaminski-Cachopo, A.; Consonni, V. Physical Properties of Annealed ZnO Nanowire/CuSCN Heterojunctions for Self-Powered UV Photodetectors. *ACS Appl. Mater. Interfaces* **2015**, *7*, 5820–5829. [CrossRef] [PubMed]

124. Lin, P.; Yan, X.; Zhang, Z.; Shen, Y.; Zhao, Y.; Bai, Z.; Zhang, Y. Self-Powered UV Photosensor Based on PEDOT:PSS/ZnO Micro/Nanowire with Strain-Modulated Photoresponse. *ACS Appl. Mater. Interfaces* **2013**, *5*, 3671–3676. [CrossRef] [PubMed]

125. Ni, P.-N.; Shan, C.-X.; Wang, S.-P.; Liu, X.-Y.; Shen, D.-Z. Self-powered spectrum-selective photodetectors fabricated from n-ZnO/p-NiO core-shell nanowire arrays. *J. Mater. Chem. C* **2013**, *1*, 4445–4449. [CrossRef]

126. Shen, Y.; Yan, X.; Bai, Z.; Zheng, X.; Sun, Y.; Liu, Y.; Lin, P.; Chen, X.; Zhang, Y. A self-powered ultraviolet photodetector based on solution-processed p-NiO/n-ZnO nanorod array heterojunction. *RSC Adv.* **2015**, *5*, 5976–5981. [CrossRef]

127. Chen, Z.; Li, B.; Mo, X.; Li, S.; Wen, J.; Lei, H.; Zhu, Z.; Yang, G.; Gui, P.; Yao, F.; et al. Self-powered narrowband p-NiO/n-ZnO nanowire ultraviolet photodetector with interface modification of Al_2O_3. *Appl. Phys. Lett.* **2017**, *110*, 123504. [CrossRef]

128. Yu, J.; Chen, X.; Wang, Y.; Zhou, H.; Xue, M.; Xu, Y.; Li, Z.; Ye, C.; Zhang, J.; van Aken, P.A.; et al. A high-performance self-powered broadband photodetector based on a CH3NH3PbI3 perovskite/ZnO nanorod array heterostructure. *J. Mater. Chem. C* **2016**, *4*, 7302–7308. [CrossRef]

129. Bai, Z.; Zhang, Y. Self-powered UV–visible photodetectors based on ZnO/Cu2O nanowire/electrolyte heterojunctions. *J. Alloy. Compd.* **2016**, *675*, 325–330. [CrossRef]

130. Guo, Z.; Zhou, L.; Tang, Y.; Li, L.; Zhang, Z.; Yang, H.; Ma, H.; Nathan, A.; Zhao, D. Surface/Interface Carrier-Transport Modulation for Constructing Photon-Alternative Ultraviolet Detectors Based on Self-Bending-Assembled ZnO Nanowires. *ACS Appl. Mater. Interfaces* **2017**, *9*, 31042–31053. [CrossRef] [PubMed]

131. Lao, C.S.; Park, M.-C.; Kuang, Q.; Deng, Y.; Sood, A.K.; Polla, D.L.; Wang, Z.L. Giant Enhancement in UV Response of ZnO Nanobelts by Polymer Surface-Functionalization. *J. Am. Chem. Soc.* **2007**, *129*, 12096–12097. [CrossRef] [PubMed]

132. Liu, K.; Sakurai, M.; Liao, M.; Aono, M. Giant Improvement of the Performance of ZnO Nanowire Photodetectors by Au Nanoparticles. *J. Phys. Chem. C* **2010**, *114*, 19835–19839. [CrossRef]

133. Gogurla, N.; Sinha, A.K.; Santra, S.; Manna, S.; Ray, S.K. Multifunctional Au-ZnO Plasmonic Nanostructures for Enhanced UV Photodetector and Room Temperature NO Sensing Devices. *Sci. Rep.* **2014**, *4*, 6483. [CrossRef] [PubMed]

134. Zeng, Y.; Pan, X.; Lu, B.; Ye, Z. Fabrication of flexible self-powered UV detectors based on ZnO nanowires and the enhancement by the decoration of Ag nanoparticles. *RSC Adv.* **2016**, *6*, 31316–31322. [CrossRef]

135. Tzeng, S.-K.; Hon, M.-H.; Leu, I.-C. Improving the Performance of a Zinc Oxide Nanowire Ultraviolet Photodetector by Adding Silver Nanoparticles. *J. Electrochem. Soc.* **2012**, *159*, H440–H443. [CrossRef]

136. Zhao, X.; Wang, F.; Shi, L.; Wang, Y.; Zhao, H.; Zhao, D. Performance enhancement in ZnO nanowire based double Schottky-barrier photodetector by applying optimized Ag nanoparticles. *RSC Adv.* **2016**, *6*, 4634–4639. [CrossRef]

137. Lin, D.; Wu, H.; Zhang, W.; Li, H.; Pan, W. Enhanced UV photoresponse from heterostructured Ag–ZnO nanowires. *Appl. Phys. Lett.* **2009**, *94*, 172103. [CrossRef]

138. West, P.R.; Ishii, S.; Naik, G.V.; Emani, N.K.; Shalaev, V.M.; Boltasseva, A. Searching for better plasmonic materials. *Laser Photonics Rev.* **2010**, *4*, 795–808. [CrossRef]

139. Lu, J.; Xu, C.; Dai, J.; Li, J.; Wang, Y.; Lin, Y.; Li, P. Improved UV photoresponse of ZnO nanorod arrays by resonant coupling with surface plasmons of Al nanoparticles. *Nanoscale* **2015**, *7*, 3396–3403. [CrossRef] [PubMed]

140. Deka Boruah, B.; Misra, A. Effect of Magnetic Field on Photoresponse of Cobalt Integrated Zinc Oxide Nanorods. *ACS Appl. Mater. Interfaces* **2016**, *8*, 4771–4780. [CrossRef] [PubMed]

141. Su, Y.; Wu, Z.; Wu, X.; Long, Y.; Zhang, H.; Xie, G.; Du, X.; Tai, H.; Jiang, Y. Enhancing responsivity of ZnO nanowire based photodetectors by piezo-phototronic effect. *Sens. Actuators A Phys.* **2016**, *241*, 169–175. [CrossRef]

142. Lu, S.; Qi, J.; Liu, S.; Zhang, Z.; Wang, Z.; Lin, P.; Liao, Q.; Liang, Q.; Zhang, Y. Piezotronic Interface Engineering on ZnO/Au-Based Schottky Junction for Enhanced Photoresponse of a Flexible Self-Powered UV Detector. *ACS Appl. Mater. Interfaces* **2014**, *6*, 14116–14122. [CrossRef] [PubMed]
143. Zhang, Z.; Liao, Q.; Yu, Y.; Wang, X.; Zhang, Y. Enhanced photoresponse of ZnO nanorods-based self-powered photodetector by piezotronic interface engineering. *Nano Energy* **2014**, *9*, 237–244. [CrossRef]
144. Zhang, F.; Ding, Y.; Zhang, Y.; Zhang, X.L.; Wang, Z.L. Piezo-phototronic Effect Enhanced Visible and Ultraviolet Photodetection Using a ZnO-CdS Core-Shell Micro/nanowire. *ACS Nano* **2012**, *6*, 9229–9236. [CrossRef] [PubMed]
145. Liu, Y.; Yang, Q.; Zhang, Y.; Yang, Z.; Wang, Z.L. Nanowire Piezo-phototronic Photodetector: Theory and Experimental Design. *Adv. Mater.* **2012**, *24*, 1410–1417. [CrossRef] [PubMed]
146. Yang, Q.; Guo, X.; Wang, W.; Zhang, Y.; Xu, S.; Lien, D.H.; Wang, Z.L. Enhancing Sensitivity of a Single ZnO Micro-/Nanowire Photodetector by Piezo-phototronic Effect. *ACS Nano* **2010**, *4*, 6285–6291. [CrossRef] [PubMed]

 © 2018 by the authors. Licensee MDPI, Basel, Switzerland. This article is an open access article distributed under the terms and conditions of the Creative Commons Attribution (CC BY) license (http://creativecommons.org/licenses/by/4.0/).

MDPI
St. Alban-Anlage 66
4052 Basel
Switzerland
Tel. +41 61 683 77 34
Fax +41 61 302 89 18
www.mdpi.com

Crystals Editorial Office
E-mail: crystals@mdpi.com
www.mdpi.com/journal/crystals

www.ingramcontent.com/pod-product-compliance
Lightning Source LLC
LaVergne TN
LVHW070621100526
838202LV00012B/701